# Early Contractor Involvement in Building Procurement

# Early Contractor Involvement in Building Procurement

Contracts, Partnering and Project Management

David Mosey

A John Wiley & Sons, Ltd., Publication

This edition first published 2009
© 2009 by David Mosey

Blackwell Publishing was acquired by John Wiley & Sons in February 2007. Blackwell's publishing programme has been merged with Wiley's global Scientific, Technical and Medical business to form Wiley-Blackwell.

*Registered office*
John Wiley & Sons Ltd, The Atrium, Southern Gate, Chichester, West Sussex PO19 8SQ, United Kingdom

*Editorial offices*
9600 Garsington Road, Oxford OX4 2DQ, United Kingdom
2121 State Avenue, Ames, Iowa 50014-8300, USA

For details of our global editorial offices, for customer services and for information about how to apply for permission to reuse the copyright material in this book please see our website at www.wiley.com/wiley-blackwell.

The right of the author to be identified as the author of this work has been asserted in accordance with the Copyright, Designs and Patents Act 1988.

All rights reserved. No part of this publication may be reproduced, stored in a retrieval system, or transmitted, in any form or by any means, electronic, mechanical, photocopying, recording or otherwise, except as permitted by the UK Copyright, Designs and Patents Act 1988, without the prior permission of the publisher.

Wiley also publishes its books in a variety of electronic formats. Some content that appears in print may not be available in electronic books.

Designations used by companies to distinguish their products are often claimed as trademarks. All brand names and product names used in this book are trade names, service marks, trademarks or registered trademarks of their respective owners. The publisher is not associated with any product or vendor mentioned in this book. This publication is designed to provide accurate and authoritative information in regard to the subject matter covered. It is sold on the understanding that the publisher is not engaged in rendering professional services. If professional advice or other expert assistance is required, the services of a competent professional should be sought.

*Library of Congress Cataloging-in-Publication Data*
Mosey, David, 1954–
 Early contractor involvement in building procurement : contracts, partnering, and project management / David Mosey.
    p. cm.
 Includes bibliographical references and index.
 ISBN 978-1-4051-9645-1 (hardback : alk. paper)  1. Building–Planning.
2. Building–Superintendence.  3. Subcontractors–Selection and appointment.
 4. Industrial procurement.  5. Construction contracts.  I. Title.
  TH438.M62 2009
  692–dc22
         2009006625

A catalogue record for this book is available from the British Library.

Set in 10 on 12 pt Palatino by SNP Best-set Typesetter Ltd., Hong Kong
Printed in the UK

1  2009

# Contents

*Introduction – Who Remembers Mott the Hoople?*     1

**Chapter 1    Early Contractor Involvement – An Overview**     6

| | | |
|---|---|---|
| 1.1 | Early contractor involvement – why bother? | 6 |
| 1.2 | Early contractor involvement and project pricing | 8 |
| 1.3 | Early contractor involvement and risk transfer | 10 |
| 1.4 | Early contractor involvement and payment | 11 |
| 1.5 | The role of building contracts | 12 |
| 1.6 | The limits of construction phase building contracts | 15 |
| 1.7 | Contractor contributions to design, pricing and risk management | 15 |
| 1.8 | The client, communications and project programmes | 16 |
| 1.9 | But do you need an early building contract? | 18 |
| 1.10 | Preconstruction commitments under framework agreements | 18 |
| 1.11 | The influence of project managers and project partnering | 19 |
| 1.12 | So what is stopping us? | 20 |
| 1.13 | Government and industry support | 21 |

**Chapter 2    Conditional Contracts and Early Project Processes**     22

| | | |
|---|---|---|
| 2.1 | The conditional preconstruction phase agreement | 22 |
| 2.2 | Recognised categories of contract | 25 |
| 2.3 | Effect of the number of parties | 27 |
| 2.4 | The planning function of contracts | 29 |
| 2.5 | Choices and contractual conditionality | 31 |
| 2.6 | Limited efficiency caused by unknown items | 32 |
| 2.7 | Risk and fear of opportunism | 33 |
| 2.8 | Conditional relationships without full consideration | 34 |
| 2.9 | Need for long-term relations | 35 |
| 2.10 | Alignment of different interests | 36 |

| | | |
|---|---|---|
| 2.11 | Preconstruction phase agreements as project management tools | 38 |
| 2.12 | Contracts and new procurement systems | 39 |

### Chapter 3  Problems and Disputes Under Construction Phase Building Contracts — 41

| | | |
|---|---|---|
| 3.1 | Introduction | 41 |
| 3.2 | The role of standard form building contracts | 41 |
| 3.3 | Standard forms and the assumption of complete information | 42 |
| 3.4 | Origins of standard forms and lack of trust | 43 |
| 3.5 | Criticisms of effectiveness of standard forms | 45 |
| 3.6 | Causes of claims | 46 |
| 3.7 | Links between claims and preconstruction phase activities | 48 |
| 3.8 | Links between claims and building contracts | 51 |
| 3.9 | New procurement procedures or gambling on incomplete information | 54 |

### Chapter 4  Early Contractor Involvement in Design, Pricing and Risk Management — 57

| | | |
|---|---|---|
| 4.1 | Introduction | 57 |
| 4.2 | Preconstruction design processes | 59 |
| 4.3 | Preconstruction pricing processes | 68 |
| 4.4 | Preconstruction risk management processes | 78 |
| 4.5 | Preconstruction subcontractor appointments | 86 |
| 4.6 | Perceived benefits of early contractor appointments | 91 |
| 4.7 | Early contractor appointments and sustainability | 92 |

### Chapter 5  Client Leadership, Communication Systems and Binding Programmes — 95

| | | |
|---|---|---|
| 5.1 | Introduction | 95 |
| 5.2 | The role of the client | 95 |
| 5.3 | The role of communication systems | 102 |
| 5.4 | The role of binding programmes | 111 |

### Chapter 6  Contractual and Non-contractual Preconstruction Options — 125

| | | |
|---|---|---|
| 6.1 | Introduction | 125 |
| 6.2 | Building contract options | 125 |
| 6.3 | Letters of intent | 130 |
| 6.4 | Non-binding arrangements | 132 |
| 6.5 | Benefits of contractual clarity | 137 |

## Chapter 7 Preconstruction Commitments Under Framework Agreements — 139

- 7.1 Commercial attraction of frameworks — 139
- 7.2 The relationship between frameworks and partnering — 141
- 7.3 Frameworks and preconstruction phase processes — 142
- 7.4 Published forms of framework agreement — 143
- 7.5 Frameworks and the Private Finance Initiative — 145
- 7.6 The impact of frameworks on changing behaviour — 146

## Chapter 8 Project Management and Project Partnering — 151

- 8.1 Introduction — 151
- 8.2 Preconstruction phase agreements and project management — 151
- 8.3 Preconstruction phase agreements and partnering — 160

## Chapter 9 Obstacles to Early Contractor Appointments — 176

- 9.1 Introduction — 176
- 9.2 Project-specific obstacles — 176
- 9.3 Procedural obstacles — 181
- 9.4 Personal obstacles — 186
- 9.5 Education and training — 193
- 9.6 The role of the partnering adviser — 195

## Chapter 10 Government and Industry Views and Experience — 197

- 10.1 Introduction — 197
- 10.2 Importance of government and construction industry support — 197
- 10.3 Support for contractor and subcontractor design contributions — 198
- 10.4 Support for two-stage pricing — 199
- 10.5 Support for selecting contractors by value — 200
- 10.6 Support for joint risk management — 202
- 10.7 Support for greater client involvement — 203
- 10.8 Government and industry views on partnering — 204
- 10.9 Government views on preconstruction phase agreements — 205
- 10.10 Industry experience of preconstruction phase agreements — 208
- 10.11 Preconstruction phase agreements in an economic downturn — 209

| | | |
|---|---|---|
| **Chapter 11** | **Conclusions – the Golden Age Of Rock 'n' Roll?** | **214** |
| 11.1 | Functions of building contracts and their potential to govern project processes | 215 |
| 11.2 | The assumption in standard form building contracts of complete project information and consequent problems and disputes | 216 |
| 11.3 | Preconstruction phase processes that can be improved by early contractor appointments | 217 |
| 11.4 | The role of the client, communication and programming of preconstruction phase processes | 218 |
| 11.5 | Contractual and non-contractual options to govern preconstruction phase processes | 219 |
| 11.6 | Increased preconstruction commitments under framework agreements | 220 |
| 11.7 | The relationship of preconstruction phase agreements to project management and partnering | 220 |
| 11.8 | Circumstances and attitudes that are obstacles to early contractor appointments or to preconstruction phase agreements | 221 |
| 11.9 | Government and industry support for early contractor appointments under preconstruction phase agreements | 222 |
| **Appendix A** | **Project Case Studies** | **225** |
| | Use of Preconstruction Phase Agreements on Single Projects | 225 |
| | Use of Preconstruction Phase Agreements on Multi-project Frameworks | 251 |
| **Appendix B** | **Preconstruction Phase Processes Under Standard Form Building Contracts** | **262** |
| 1 | Introduction | 262 |
| 2 | Design development | 263 |
| 3 | Two-stage pricing | 269 |
| 4 | Risk management | 271 |
| 5 | Communications | 275 |
| 6 | Programmes | 283 |
| 7 | Team integration | 288 |
| **Appendix C** | **Preconstruction Phase Processes Under Standard Form Framework Agreements** | **291** |
| | Introduction | 291 |
| | JCT Framework Agreements | 291 |
| | NEC3 Framework Contract | 294 |
| **Appendix D** | **Form of Risk Register** | **298** |

| Appendix E | Form of Partnering Timetable | **299** |
| Appendix F | Association of Partnering Advisers Code of Conduct | **300** |
| Appendix G | **Bibliography** | **302** |
| | Part 1 Contract forms | 302 |
| | Part 2 Articles and books | 304 |
| | Part 3 Government/industry reports | 306 |
| | Part 4 Table of cases | 308 |

Index                                                                 309

# INTRODUCTION
## Who Remembers Mott the Hoople?

When speaking in public on the potential for change in construction procurement, I would often ask delegates to raise their hands in response to the following three questions:

- Who has worked on a partnering project?
- Who has worked on a partnering project with early contractor involvement?
- Who remembers the rock band 'Mott the Hoople'?

As this band had its hits in the early 1970s, I suggested that those who kept their hands up for all three questions were living proof that you can teach an old dog new tricks. This worked well until I was cornered in Kensington Town Hall by a diehard Mott the Hoople fan who talked me through every gig he had ever attended, while I watched my chances to speak with other delegates evaporate.

Some years on, I remain concerned by the difficulties that many in the construction industry seem to have in mastering and applying the trick of early contractor involvement. It was this concern that gave rise to the research on which this book is based, linked to my conviction that early contractor involvement is fundamental to successful partnering and that both require the support of new forms of building contract.

I have to admit from the outset that I like building contracts and that, while I appreciate they may not be everyone's favourite read, I believe that this is often because they fail to fulfil their potential. For example, what is the problem with building contracts that so many project teams say they want to 'keep them in the drawer'? Is this phrase really just a euphemism for hoping to avoid disputes? Or does it reveal an assumption that the agreement and conditions forming part of the building contract (as distinct from the drawings, specifications and pricing documents) do not contribute to the success of a project except as a means of resolving disputes?

Clearly, the creation of a robust agreement that protects rights and limits liabilities is of fundamental importance in a sector such as construction, where risks are high and margins are limited, but is this the only role that building contracts can fulfil? Could they not also guide and support the project team, particularly if put in place at an earlier stage?

The preparatory processes for building and civil engineering projects take a lot of time and are the subject of considerable client investment in design consultants' fees[1]. However, the structured involvement in these processes of the main contractor who will build the project, and of its subcontractors and suppliers with a design capability, is often very limited or non-existent[2]. If the building contract agreement and conditions govern only the construction phase of the project, then it is fair to say that by the time they are signed up all the creative stages of the project are largely complete and everyone primarily wants to get the job finished with minimum hassle. Yet when disputes do arise and the building contract comes out of the drawer, the origins of the disputes are often found to be in the early stages of the project that the construction phase building contract has done nothing to influence[3].

Against this background is there merit in creating earlier conditional building contracts to govern an earlier role for contractors in all or part of the preconstruction phase? Could this alternative model help to tackle the causes of disputes, and could an earlier building contract be used to require or encourage efficient procurement and project management practices?

Over recent years, clients have worked more closely with contractors through a team-based approach to projects known as 'partnering'[4]. Yet despite a wealth of successful projects and widespread government and industry endorsements, there remains persistent confusion as to what partnering actually means and as to what it requires from project team members. This has slowed its progress and has allowed the cynics to suggest that it is at best no more than hot air, and at worst a dangerous means of engagement between clients, consultants and contractors that is open to exploitation at the first sign of trouble[5].

Is it possible that building contracts can help to resolve the paradox whereby partnering is widely supported and appears to work but still lacks a clear, consistent definition? Can a team describe in writing the

---

[1] Under standard form appointments, RIBA (2004) and ACE (2002), 75% of an architect's or engineer's fees are payable during the preconstruction phase.
[2] RIBA (2004) does not mention any main contractor, subcontractor or supplier involvement in preconstruction design development and related processes, and RIBA (2008) sees even the design and build contractor as a client rather than a contributor. See also Chapter 4 Section 4.2.2 (Integration with consultant designs).
[3] For example, inaccurate design, inadequate design and inadequate site investigation: Kumaraswamy (1997), 21–34. See also Chapter 3 Section 3.6 (Causes of claims).
[4] See Chapter 8, Section 8.3.1 (What is partnering?) for definitions of partnering.
[5] See Chapter 8, Section 8.3.4 (Challenges to successful partnering).

## Introduction

key features of their partnering relationship without losing its magic? Is it possible for contracts to underpin those relationships by mapping out the processes through which they are built up?

The creation of conditional early contractor appointments, containing agreed team-based processes and programmes, can greatly assist the establishment of successful partnering relationships. This requires on the one hand an adjustment in the traditional view of what a contract can achieve, and on the other hand acceptance of more prosaic wording than some partnering enthusiasts will find entirely satisfying. However, when the links are properly made between early contractor appointments and partnering, the benefits to the client and the construction industry can be significant.

This book has its origins in the drive for reform of construction procurement which followed the Egan Report in 1998[6]. The radical rethink of construction procurement recommended by Egan led to my participation in a task-force that produced the Construction Industry Council (CIC) guide to project team partnering[7], my authorship of the PPC2000 form of partnering contract[8] and my involvement as adviser on a large number and variety of partnering projects. At the time when PPC2000 was published, building contracts had been criticised by Egan as an obstacle to good construction practice[9], but were recognised by the CIC as a necessary medium for disseminating and embedding successful new approaches to project procurement[10].

Based on a template produced by the CIC for a partnering contract form[11], PPC2000 described a two-stage procurement and contractual model that provided for the early conditional appointment of the main contractor (and potentially certain of its subcontractors) alongside the client's consultants during the preconstruction phase of the project. An illustration of this two-stage model is set out in Project Flowchart 2 at the end of Chapter 3. There is a wealth of evidence to suggest that such an early conditional contractor appointment can bring significant benefits to all parties engaged on the relevant project. However, although several other building contract forms have been introduced or overhauled since the publication of PPC2000, including some that address preconstruction services, none of them to date have adopted an equivalent two-stage approach to early contractor involvement.

---

[6] Egan (1998).
[7] CIC (2002), originally published June 2000.
[8] PPC2000, published in September 2000.
[9] 'Contracts can add significantly to the cost of a project and often add no value for the client', Egan (1998), Section 69, 33.
[10] 'An effective contract should support the full partnering team and aim to deliver an integrated project process' CIC (2002), 12.
[11] As set out in CIC (2002), 14–23.

The single-stage procurement and contractual model, whereby the main contractor and its subcontractors are selected and appointed only for the construction phase of a project, remains the dominant approach. An illustration of this single-stage model is set out in Project Flowchart 1 at the end of Chapter 3. The reasons for the enduring use of this model may be attributable to its familiarity, or to its simplicity, or to other economic, procedural or cultural factors. However, it is arguable that such a model does not necessarily obtain the best contributions of all parties to a successful project as it omits the main contractor and subcontractors from all early design and project planning. Is it therefore worth examining further the potential of early contractor involvement as an alternative procurement model and, for this purpose, the place of the conditional preconstruction phase agreement in contract theory, in addressing the needs of the construction industry and its clients on specific projects, and in comparison to other procurement and contracting systems?

I hope to challenge the assumptions inherent in the dominant single-stage procurement and contractual approach and to analyse whether in certain circumstances the use of a conditional preconstruction phase agreement offers benefits that reward the effort of appointing contractors at an earlier stage in the procurement process. I will question whether the apparent certainties of cost, time and quality attributed to single-stage procurement are at the expense of other important considerations and whether the protections and administrative machinery created by construction phase contract forms comprise the only roles that a building contract can fulfil. I hope to explain the ways in which a building contract can describe and support design procurement and risk management processes if it is set up initially as a conditional agreement. I will also consider the extent to which the value of such a conditional agreement is linked to or dependent on the use of the collaborative approach to project management known as partnering or to the award of successive projects to the same team under a framework arrangement.

The benefits of early contractor appointments under conditional preconstruction phase agreements will be illustrated by reference to twelve case studies of individual projects and of multiple projects under framework agreements, further details of which are set out in Appendix A. These case studies have been built up from review of relevant contract documents, from discussions and correspondence with project team members and from reports published by such bodies as the Highways Agency, Constructing Excellence and the Housing Forum[12].

Discussion of the contractual options for creating conditional preconstruction phase agreements is supplemented by, and cross-

---

[12] See Project case studies 1 to 12, Appendix A.

## Introduction

referenced to, a detailed comparison of six published standard form building contracts in Appendix B and a further comparison of two published standard form framework agreements in Appendix C.

This book is based closely on the doctoral research that I submitted to King's College, London, under the heading 'Process contracting and early contractor appointments: the potential of the conditional preconstruction phase agreement to support procurement, partnering and project management'. I would like to thank Professor Philip Capper and Professor Philip Britton for guiding my research, my wife, Cécile, for encouraging my work, and my colleagues in the Projects and Construction team at Trowers & Hamlins LLP for their cheerful commitment to innovative project procurement.

# CHAPTER ONE
# EARLY CONTRACTOR INVOLVEMENT – AN OVERVIEW

## 1.1 Early contractor involvement – why bother?

The majority of published standard form building contracts provide for the appointment of the main contractor and its subcontractors and suppliers at the point when construction is due to commence[13]. They are generally preceded by a single-stage procurement exercise to select a suitable contractor who has offered a price based on designs developed by other parties. But does this approach always reflect the wishes and needs of the industry and its clients, or does it instead reflect a long-established status quo in a complex and fragmented sector? Arguably, it is often the latter.

If there are benefits to be gained from earlier contractor involvement, is a contract necessary or even desirable to achieve this? Should a conditional building contract govern early project processes, particularly where design processes overlap with the procurement processes by which prices for those designs are agreed? Yet if contracts do not enter this territory, project teams will lack necessary guidance as to the nature and extent of a contractor's early involvement, and its attendant rights, obligations and risks.

Government reports as early as Emmerson in 1962 identified the separation of the design phase from the construction phase of the project as a problem, and observed that 'In no other important industry is the responsibility for the design so far removed from the responsibility for production'[14]. The Banwell Report in 1964 picked up this theme and stated that 'those who continue to regard design and construction as separate fields of endeavour are mistaken'[15]. Nearly 30 years later, Sir Michael Latham observed that many of the problems identified by Banwell had not been solved and that among these 'the traditional

---

[13] For example, JCT 2005 SBC/Q and JCT 2005 Design and Build; also NEC3.
[14] Emmerson (1962), 9.
[15] Banwell (1964), 4, Section 2.6.

separation of design and construction has long been a source of controversy'[16].

The client draws no distinction between design and construction when occupying the completed project, and is interested only in obtaining the benefit of a project completed efficiently without claims or disputes. Without a clear preconstruction contractual model there is a greater likelihood of decisions being delayed or sidestepped, thereby deferring main contractor and specialist appointments and perpetuating the problems of separating the design of a project from its construction.

It has long been recognised that design contributions should be made not only by consultants but also by contractors and specialist suppliers and fabricators to achieve a complete and functional design. For example, in respect of electrical systems and heating, ventilation and air conditioning ('HVAC'), Smith *et al.* found that the split of HVAC responsibility under traditional procurement methods was often unrealistic and arbitrary. They suggested that in fact HVAC design is made up as follows:

> 'The overall design … may be the responsibility of a design consultant, the co-ordination of M&E services may be the responsibility of the main contractor, while the detailed design of the HVAC installation may be the responsibility of a specialist supplier. The structural engineer meanwhile is responsible for the design of the building frame, although detailing may be the responsibility of the fabricator'[17].

Commentators have recognised that a procurement model which omits contractor and specialist design contributions can increase risk and can result in poor communications between team members, unnecessary delays to progress of the project and the creation of incorrect information that leads to claims and disputes. I will argue that the most effective way to add value and to challenge the risks of excluding contractor contributions is for clients, consultants and contractors to form a full team at an early stage in the project, establishing the roles of all parties under integrated conditional preconstruction phase agreements.

The central propositions in this book are:

(1) That a significant number of construction projects suffer from inefficiencies, claims and disputes for reasons that can be traced to the late appointment of the main contractor and key subcontractors

---

[16] Trust and Money (1993), 7.
[17] Smith *et al.* (2006), 14.

and suppliers, and to the consequent inadequacy of preparatory and planning activities that need to be undertaken during the preconstruction phase;

(2) That neglected preparatory and planning activities include insufficient involvement of the main contractor (and to some extent its subcontractors and suppliers) in joint working alongside the client and its consultants on design development, the finalisation of works and supply packages and their costs, the analysis and management of project risks and their costs, and the agreement of a construction phase programme;

(3) That conditional preconstruction phase agreements (whether bespoke or forming part of a standard form building contract) have a greater role to play in governing these preparatory and planning activities, and that such agreements, by setting out these activities as a series of interlinked processes, can operate as a valuable tool for project managers, particularly if they are subject to agreed preconstruction phase programmes and to the systems of open communication and collaboration known as 'partnering'.

## 1.2 Early contractor involvement and project pricing

One commercial issue that needs to be addressed from the outset is the fact that an early contractor appointment to participate in design development, risk management and construction phase programming is unlikely to be on the basis of a fixed price. If the contractor is appointed to work alongside the client and its consultants in developing additional information in these areas and in finalising an acceptable price prior to start on site, then logically there will be insufficient information available for detailed or accurate pricing to be undertaken prior to commencement of such work. It is therefore relevant to consider the implications of this in terms of criteria for early contractor selection and the means by which preconstruction phase processes involving the contractor can lead the parties to achieve the required level of cost certainty after early conditional contractor appointment, but prior to unconditional contractor appointment.

I will consider whether there are weaknesses in the system of single-stage fixed price tendering in a marketplace where main contractors obtain many of their specialist skills and supplies from subcontractors and suppliers[18]. In order to provide an accurate price in a single-stage tender, each bidding main contractor would in theory need to present the client's proposed requirements to each of its subcontractors and

---

[18] See for example Chapter 4, Section 4.3 (Preconstruction pricing processes).

suppliers so as to obtain subdivided fixed price quotes prior to each main contract bidder, then submitting its own fixed price quote to the client. The time and cost of conducting such procedures in a structured and thorough manner in preparation for every tender are prohibitive for most main contractors, subcontractors and suppliers on most projects.

This is due in part to time constraints set by the client for the main contract tender process and in part to the cost and difficulty for the main contractor of subdividing the client's tender documentation so as to obtain separate quotes from each subcontractor and supplier. This practical challenge is exacerbated by the large number of tenders sent out by clients to prospective main contractors and the even greater number of subdivided subcontract tenders that would have to be sent out by each tendering main contractor to a range of prospective subcontractors and suppliers. The resources required for subcontract tenderers to compile their bids with accuracy, even if the main contract tender period was sufficiently long, would give rise to considerable costs. Hence, main contractors and their subcontractors and suppliers are likely to make judgements as to the level of detail and accuracy required in their enquiry documents and responses, according to the importance of each project element, and to allow additional amounts to cover the risk of inaccurate pricing.

It has been recognised that any contractor will be at risk if it is obliged to provide a fixed price quotation to a client based only on budget estimates received from its subcontractors when those subcontractors are not in a position themselves to give a fixed quotation, for example because suitably detailed drawings are not made available[19]. In single-stage fixed price tendering, bidding contractors may not be allowed the opportunity to comment on whether the designs forming part of the invitation to tender are sufficiently detailed for them to obtain fixed price quotations from their subcontractors and suppliers sufficient to compile an accurate total price. Although many clients and consultants remain uncomfortable with the appointment of a main contractor in advance of agreeing a fixed price, I will challenge whether a fixed price quote obtained at arm's length is likely to be accurate or reliable except in limited circumstances. I will also explore the means by which a conditional preconstruction phase agreement can offer the client ways to achieve better control over costs, for example through open-book agreement of profit, through joint evaluation and approval of supply chain prices and other cost components, and through incentives for contractor and consultants to bring costs down at all stages of the project.

---

[19] See for example Burke (2002), 85, as to the risks of single-stage pricing.

## 1.3 Early contractor involvement and risk transfer

Another commercial issue that needs to be tackled is the fact that, as additional information is built up following an early contractor appointment, it will not be possible for the client to transfer risks that emerge later in the preconstruction phase of the project if the contractor is not willing to accept them.

Risk management is not an orderly sequential process comparable to other preconstruction activities and it is not possible to guarantee in advance that joint risk management involving the contractor will lead to a risk and cost position acceptable to all parties. However, for the client and its advisers to seek fixed prices from a main contractor without recognising the scope for it to contribute to early risk management is to draw a veil over important commercial factors. Specifically, under a traditional single-stage contractor appointment, if risks arise during construction which the main contractor has not foreseen at the time of its tender or if a risk contingency allowed by the main contractor proves to be insufficient, it is unlikely that the main contractor will allow a profitable job to become loss-making simply because it accepted those risks within its fixed price. Instead, this situation is likely to give rise to manoeuvring and claims by the main contractor to try to recoup any loss deriving from its miscalculation. This in turn can be prejudicial to the quality of the project, for example if the main contractor looks for ways of cutting costs that may not be in the interests of the client and may not be declared to the client.

In one case study of successful risk management, it was noted that the risks allocated to the contractor were those that it was able to manage[20]. Such risks should not be allocated on the basis of expediency, which can be the result of a priced-based single-stage tender. In the long run it is better value for a client to pay for risks that actually occur during the construction phase of a project rather than to agree a price based on what a contractor thinks might occur. In the latter case, risk is transferred arbitrarily and both the client and the contractor are gambling on whether that risk has been accurately costed.

Where contracts continue to focus only on the transfer of risk and not on its management, it has been observed that this will usually give rise to a risk premium charged by the party accepting the transferred risk. It is possible that the risk premium charged by the contractor (or by a subcontractor or supplier) is insufficient to cover the cost of the required remedial action if and when that risk materialises. In those circumstances, it is likely that the contractor, subcontractor or supplier will be unwilling to incur the additional costs necessary to cover the

---

[20] See the case study of successful risk management described by Smith *et al.* (2006) at 75, 76.

risk as this will erode its profit. As a result, the project will suffer from the claims and counter-claims that arise as the client seeks to impose the risk transfer provision and as the contractor, subcontractor or supplier seeks to resist incurring costs that make the project unprofitable. In these circumstances, the client and its project are likely to suffer more adverse consequences than the cost of the client retaining the risk or agreeing a joint strategy with the contractor for managing it[21]. I will therefore consider whether the early appointment of a contractor, so that it participates in joint risk management with the client and consultants, can give rise to tangible benefits. For example joint risk management may avoid or reduce contractor risk premiums normally invisible to the client, but nevertheless payable under single-stage construction phase contracts that have no such facility.

## 1.4 Early contractor involvement and payment

A third commercial issue that is fundamental to early contractor appointment is money. How should the contractor be remunerated for the activities that it undertakes during the preconstruction phase?

If clients hope to obtain contractor contributions at no cost, they risk first, the contractor not applying sufficient resources to the required tasks, and second, a loss of contractor objectivity and professionalism[22]. A 1975 National Economic Development Office (NEDO) Report contained a county borough housing case study that recognised cause for concern where early client/main contractor collaborative working can tempt the client to seek design enhancements without recognising their cost consequences. It found that involvement of the main contractor at the design stage 'led to numerous small disputes on detail where the client wanted more expensive solutions to the specification' and where the resultant increased costs 'fall entirely on the contractor's profit margin if the price is already fixed'. It is interesting that in this case study it was assumed that early contractor involvement in design could somehow be achieved after a fixed price had already been established, inferring that the contractor would make a price commitment on limited design information and would then be expected to take the cost risk of client enhancements that emerged in later detailed design development. This seems commercially unrealistic and unlikely to generate successful joint working. It also fails to recognise the benefits identified by Banwell of appointing the contractor to work as part of the team, not only in finalising the details of the project but also in establishing its cost[23].

---

[21] See for example Smith *et al.* (2006), 62, 63, as to premiums charged for the arbitrary transfer of risk.
[22] NEDO (1975), 116, Table B1.
[23] See also Chapter 4, Section 4.3.4 (Prices and contractor selection).

If the main contractor will only be paid if the project goes ahead, then surely commercial logic dictates that its first priority is likely to be ensuring that the project goes ahead whatever the cost to the client. Also, where construction phase profit and consultant fees are calculated as a percentage of cost, a cynic might suggest that contractor's interests and those of the consultants are best served by persuading the client to build its project on the largest possible scale, particularly if they can expect no other reward for offering the client cheaper options through value management or value engineering that would have the effect of reducing their percentage take.

A 1998 report by the Construction Industry Research and Information Association (CIRIA) stated that 'contractors must be appropriately rewarded for contributions made to the project', and also that project contracts should be structured so as to 'recognise all the contributions being made, and the related risks, responsibilities and rewards, particularly during project development'[24]. The Housing Forum (2000) report *How to Survive Partnering – It Won't Bite* recorded in its survey findings the suggestion from client/consultant/contractor respondents 'that the contractor should be paid as a consultant during the lead-in period before the contract is signed' (i.e. during the preconstruction phase)[25]. A 2005 National Audit Office (NAO) report pointed to case studies that include a Milton Keynes Treatment Centre where 'For three months, the principal supply chain partner worked on a fee basis, developing options for the hospital to consider'[26].

In order for the client to obtain added value from a contractor's preconstruction phase contributions, is it not preferable for the contractor to join the consultants as part of the professional team and to provide its early services for an appropriate reward? I will argue that this is more likely to secure value for the client than an extended period of speculative endeavour where any contractor reward is entirely contingent on the construction phase of the project proceeding.

## 1.5  *The role of building contracts*

Building contracts, like other contracts, assume differing commercial interests of the parties who enter into them, and need to protect those interests while also reconciling them through the prospect of agreed payments. R.J. Smith identifies three roles for building contracts:

- To set out rights, responsibilities and procedures;
- To identify, assign and transfer risk;

---

[24] CIRIA (1998), 15.
[25] Housing Forum (2000), 13.
[26] NAO (2005) Case Studies, 33.

*Early Contractor Involvement – An Overview*

- To act as a 'planning tool' so that there are 'fewer surprises and dilemmas during construction'[27].

By the time that work commences on site, all the planning has already been done. Hence, in order to be an effective planning tool, a building contract needs to exist at the time when the parties are doing the planning and to describe the systems by which the planning takes place. Yet the earlier it is created in the project processes, the more conditional its terms are likely to be.

N.J. Smith described a building contract as a means to 'formalise a set of risks, rules and relationships into one set of words which will govern all dealings between the parties while carrying out that contract'[28]. If there is a need for preconstruction phase dealings between the client and main contractor, clearly a building contract will not govern all dealings if it omits that phase.

Standard form building contracts have evolved to reflect changes in procurement practices[29]. This book will examine what is arguably the latest stage of that evolution, namely the role of the building contract and in particular the preconstruction phase agreement as a 'procurement system'[30], assisting the client, consultants, main contractor and specialist subcontractors in moving from incomplete to complete information through interrelated design processes, procurement processes, risk management processes and programming processes.

A practical illustration of the move towards a greater focus on preconstruction phase activities is given by the National Audit Office in relation to development of the procurement and contractual systems used by the University of Cambridge[31]:

Such evolution may justify a shift away from the unconditional and transactional functions of a building contract towards reliance on planning systems set out in the contract itself. It will be argued that a conditional preconstruction agreement as a procurement system has many of the features of a 'neo-classical' contract, an incomplete

---

[27] Smith (1995), 41 and 42.
[28] Smith (2002), 178.
[29] For example, the Joint Contracts Tribunal (JCT) published its first design and build contract in 1981 (JCT WCD), its first management contract in 1986 (JCT Management Contract), its first construction management suite of contracts in 2002 (JCT CM) and its first partnering contract in 2007 (JCT CE).
[30] Arup, in their report to the Office of Government Commerce (OGC), describe PPC2000 as 'a procurement system that provides the processes and mechanisms for planning, procurement and delivery of construction works. The system is based on the application of a number of processes and it is essential that the processes stated are applied.' Arup, 2008, 37.
[31] NAO (2005) Case Studies, 15.

| NAO (2005) University of Cambridge Case Studies ||| 
|---|---|---|
| **Contract 1 (1998)** | **Contract 2 (2000)** | **Contract 3 (2002)** |
| Traditional single-stage tender awarded on lowest price. No contractor involvement in design. Cost and time overruns, with buildings containing many defects and relationships with the contractor strained.<br><br>**Cost: +2%**<br>**Time: eight weeks late**<br>**Client satisfaction: 6/10**<br>**(post-project completion review six months after practical completion)** | Two-stage tendering process (JCT98 contract). Contractor involved in design. Effective teamwork. New contractor, so limited lessons learnt from repeat work.<br><br>**Cost: +0%**<br>**Time: on time**<br>**Client satisfaction: 7/10** | Two-stage contract (New Engineering Contract), with a professional services contract used for the first stage, and the contractor and principal subcontractor involved in the design. Selection on transparent criteria (30% quality: 70% price balance), with the original contractor re-engaged and so lessons brought to bear along with effective teamwork. Changed user move dates successfully met.<br><br>**Cost: −3%**<br>**Time: On time**<br>**Client satisfaction: 9/10** |

agreement containing machinery for dealing with matters that remain to be resolved between the parties[32].

In Chapter 2, I will examine the features and contractual types that make up a conditional preconstruction phase agreement and the features of contract law that may affect its efficiency as a contract, including the extent to which the parties are dependent on their relationship as well as their specific contractual obligations. I will consider whether there is a risk of a conditional preconstruction phase agreement being unenforceable for reasons of uncertainty or incompleteness and whether such an agreement can offer a clear path through the 'relational' activities that are features of the approach to project management known as partnering.

---

[32] Williamson distinguishes 'classical', 'neo-classical' and 'relational' contracts, as considered further in Chapter 2, Section 2.2 (Recognised categories of contract), and notes that 'A recognition that the world is complex, that [the] agreements are incomplete, and that some contracts will never be reached unless both parties have confidence in the settlement machinery [thus] characterises neo-classical law.' Williamson (1979), 238.

## 1.6 *The limits of construction phase building contracts*

In order to make the case for a new type of contract governing two-stage procurement, it is necessary to consider the shortcomings of single-stage construction phase contracts. The construction industry makes extensive use of published standard form construction phase building contracts, many of which appear to assume the availability of complete project information at the point when they are created. Do these forms do the whole job that a contract can and should do? I suggest that they do not.

However, the investment and effort required to adopt new contract forms that deal with preconstruction processes are considerable and can only be justified if clients and other project team members risk suffering significant loss in their absence. With this in mind, what are the predominant causes of claims and disputes on building projects? Can they be traced to failures in preparatory and planning processes, and to what extent can they be addressed by a new contractual treatment of these processes?

In Chapter 3, I will explore the implications of the fixed paradigm by which the information needed to complete and implement most standard form building contracts appears to require that when the relevant project was put out to tender there was substantially complete project information available, including designs sufficient for main contractor bidders to assess all relevant project risks and quote fixed prices. It will be argued that this assumption that project information is complete prior to creation of a building contract has been the cause of misunderstandings and failures that have led to significant inefficiencies, claims and disputes. I will also suggest that the most prevalent causes of claims and disputes are directly linked to failings in preconstruction phase activities and, arguably, the absence of the main contractor from the team while such activities are being undertaken.

## 1.7 *Contractor contributions to design, pricing and risk management*

It has been suggested that the greater opportunities for improving the parties' performance and the overall project results are at the 'front end' of the project process[33]. For example, Burke observed that the ability of the parties to influence project outcomes, including reduction of cost, creation of additional value, improvement of performance and flexibility to incorporate changes is much higher in the earlier

---

[33] Burke (2002), 31.

conceptual and design stages of the project. It is evident that by the time the construction and other aspects of project implementation are underway, the ability of any party to reduce cost or implement other changes in an efficient manner has reduced significantly. Burke was arguing the case for early project manager appointments, but the same argument can equally be used to support the case for early contractor and specialist appointments.

If problems, disputes and inefficiencies arise in the absence of contractor involvement in project preparation, the next step is to consider in more detail the ways in which contractors can contribute to such preparation and whether or not their contributions add significant benefits or give rise to other problems, disputes and inefficiencies of their own.

In Chapter 4, and using eight of the project case studies set out in Appendix A, I will illustrate the ways in which early appointment of main contractors and specialist subcontractors can improve preconstruction phase processes, including design development, finalising of supply chain members and prices, and risk management[34]. The eighth project case study (Project X) will illustrate circumstances where a preconstruction phase agreement was not successful in achieving its intended purposes, identifying the reasons for this and how they could be avoided[35].

## 1.8   The client, communications and project programmes

As clients have expressed dissatisfaction with traditional models[36], it is therefore important to explore the extent to which clients have a new role to play in encouraging implementation of alternative models. To quote Latham 'Implementation begins with clients. Clients are at the core of the process and their needs must be met by the industry'[37]. In Chapter 5, I will review typical client roles under construction phase building contracts and will argue the need for closer client involvement under conditional preconstruction phase agreements.

Whether they are interested in early project processes or not, clients are generally the only common signatory to a series of two-party consultant appointments and a building contract[38]. Hence, at some stage

---

[34] See Chapters 4 and 5 and Project case studies 1 to 7, Appendix A.
[35] See Chapters 4 and 5 and Project case study 8, Appendix A.
[36] See Chapter 3, Section 3.5 (Criticism of effectiveness of standard forms).
[37] Latham (1994), 3.
[38] Exceptions to this are multi-party contracts such as PPC2000 under which the consultants, main contractor and certain subcontractors and suppliers are also in direct contractual relationships with each other. See also Appendix B.

the client is likely to be asked to make project decisions based on conflicting views and information provided by different team members in their respective roles and applying the terms of their different contracts. Chapter 5 will consider whether closer involvement earlier in the project can assist the client in dealing with such situations and will illustrate by reference to project case studies the ways in which a greater client involvement may benefit particular project processes.

For a project team to work efficiently, communication is required between individuals as well as organisations. Where they are tasked with important project decisions, such individuals need guidance as to who has what authority, when they should meet, how they should reach decisions and what contractual effect those decisions will have. I will explore how in a conditional preconstruction phase agreement, where a decision-making process is part of the system for moving from incomplete to complete information, an agreed contractual system of communication can be of particular importance to facilitate such decision making.

The establishment and bedding in of a communication system is itself a preparatory process for the construction phase of the project, and it will be suggested that this will operate more effectively if the agreed individuals remain the same during the preconstruction and construction phases and operate as a 'core group'[39], with a duty to provide mutual 'early warning'[40] of problems and to seek agreed solutions. The project case studies will be used illustrate how this can work in practice.

If preconstruction phase contractual commitments are to achieve benefits over and above less formal arrangements, they need to be subject to programmes to identify who does what during the preconstruction phase, and when key activities will be completed. But should these programmes be contractually binding? It will be argued that identifying and agreeing contractual deadlines for key preconstruction phase activities is central to the success of a preconstruction phase agreement.

Chapter 5 will consider the status of programmes in published standard form consultant appointments as well as in standard form building contracts, and also the sensitivities of programming the creative processes by which designs are conceived and developed. It will be argued that there is a need to agree contractually binding deadlines for all preconstruction phase activities, including design outputs, pricing exercises and the early agreement of a construction phase programme.

---

[39] The contractual status of a 'core group' was first recognised in PPC2000 and now also appears in NEC3 Option X12 and Perform 21.
[40] The contractual status of 'early warning' was first recognised in NEC2 and also appears in PPC2000 and Perform 21.

The project case studies will be used to illustrate the benefits that this approach can achieve and the problems it can avert.

## 1.9  But do you need an early building contract?

Where contractors, subcontractors and suppliers are involved in project processes ahead of start on site, it is often through an informal arrangement such as a 'letter of intent' rather than a preconstruction phase agreement. A wide range of other options have also been employed to obtain early contractor input, varying from corporate joint ventures to non-binding project protocols.

In Chapter 6, I will compare the contractual and non-contractual options for describing preconstruction phase processes and will assess the pros and cons of each. I will consider the steps necessary to create preconstruction phase appointments under the published standard form building contracts GC/Works/1, NEC3, PPC2000, Perform 21, JCT 2005 and JCT CE by reference to a comparison of these forms of contract set out in Appendix B[41]. I will contrast the use of less formal techniques such as letters of intent and non-binding protocols, and the implications of relying on personal relationships with no contractual support at all.

Although it is possible to achieve the benefits of early involvement of the main contractor and its specialist subcontractors and suppliers in preconstruction phase processes without a legally binding preconstruction phase agreement, it will be argued that such an agreement offers a clearer and better structured approach that is more likely to be understood and adhered to, and is therefore more likely to achieve the team's objectives.

## 1.10  Preconstruction commitments under framework agreements

Having examined preconstruction phase processes and relationships in relation to a single project, what differences arise if these processes and relationships are applied over a series of projects? In Chapter 7, I will look at the impact of framework agreements on the industry's approach to preconstruction phase activities, and consider whether the performance of such activities as preconditions to the unconditional award of successive projects is relevant to the success of framework agreements in practice. Appropriate features of framework agreements

---

[41] See review of standard form contracts in Appendix B.

*Early Contractor Involvement – An Overview*

are considered by reference to the standard form framework agreements published by JCT[42] and NEC[43] reviewed in Appendix C[44].

Four further project case studies will illustrate how framework agreements have been used to describe preconstruction phase processes as well as to set out the criteria for award of successive projects[45], and to identify particular benefits achieved pursuant to these framework agreements. I will also illustrate by reference to the last project case study (Project Y) the problems encountered under a framework agreement where preconstruction phase processes were not implemented, noting the reasons for these problems and suggesting how they could be avoided[46].

## 1.11 The influence of project managers and project partnering

The provisions of a preconstruction phase agreement merge contractual processes with the project management of design, procurement, risk and start up on site. It is therefore important to consider the responsibilities of the project manager, particularly during the preconstruction phase, and how fulfilment of these responsibilities could be affected by the creation of a preconstruction phase agreement. In Chapter 8, I will look at the purposes of project management and the role of the project manager during the preparatory and planning stages of a project.

I will also consider the influence of project managers over the choice of procurement strategy and their role in organising communications and the integration of the team, highlighting the need for objectivity if the project manager is to be credible to all team members. It will be argued that a strong project manager is not a substitute for a conditional preconstruction phase agreement, and that project managers should welcome the additional clarity created by such an agreement.

Chapter 8 will then review the functions and features of partnering, and will categorise it as a type of project management underpinned by teamwork between different organisations. Focusing on partnering as applied to implementation of a single project, I will argue that the collaborative activities that together comprise partnering should be undertaken throughout all stages of a project, but that they are particularly important during the preconstruction phase for the following reasons:

---

[42] JCT 2005 Framework Agreement.
[43] NEC3 Framework Contract.
[44] See review of the JCT Framework Agreement and NEC Framework Contract in Appendix C.
[45] Project case studies 9 to 11 inclusive, Appendix A.
[46] Project case study 12, Appendix A.

- This is when new relationships are being formed and there is still thinking time during which teamwork can be applied in the search for new or improved designs, sources of supply and construction techniques.
- This is when the composition of the team and their collaborative working methods can be trialled without unconditional commitment and can be altered if unsuccessful.

I will propose that the development of successful partnering relationships and working methods on a conditional basis benefit from the clarity and discipline of contractual terms. I will suggest that a conditional preconstruction phase agreement is, therefore, directly relevant to the success of project partnering, and that by describing agreed partnering processes as features of project management such an agreement can complete a missing link that otherwise separates partnering from building contracts.

I will also consider the risks that arise in partnering if the parties focus only on collaborative values and teamwork without creating team-based contractual systems to deal with problems or without agreeing deadlines to drive their project partnering processes forward. I will consider the impact of potential challenges to the success of partnering, such as varying organisational cultures, changing business conditions, uneven levels of commitment and lack of momentum and will address whether each of these challenges can be overcome through appropriate provisions set out in a preconstruction phase agreement.

## 1.12  So what is stopping us?

If conditional agreements exist that can govern preconstruction phase activities, why are they not used more widely? Are there good reasons for resistance to a two-stage contractual approach and are there particular types of project for which it is inappropriate?

In Chapter 9, I will seek to identify those types of projects that may not be suitable for the early appointment of the main contractor and its specialist subcontractors and suppliers. I will consider whether preconstruction phase agreements may also be inappropriate for other reasons, and will seek to identify where by contrast they may fail to fulfil their intended purposes because they are not properly implemented by the parties.

Reasons for resistance reviewed in this way will include project-specific, procedural, cultural and personal obstacles to the use of preconstruction phase agreements, noting on the one hand where the reasons for such resistance are logical and on the other hand where they may result from misunderstandings or from the difficulty of changing a long-established status quo until such time as decision makers are otherwise influenced by training and education.

## 1.13 Government and industry support

Notwithstanding the apparent benefits that can be obtained through the wider use of preconstruction phase agreements, and in addition to training and education as to the way they can work, sustained and influential support will be required for such agreements to be carried into the wider construction marketplace.

In Chapter 10, I will review how the future increased use of such agreements may be influenced by powerful groups of public and private sector clients and government and industry best practice bodies according to the extent that they identify, encourage, implement and benefit from early contractor, subcontractor and supplier appointments as a means to ensure improved project processes. In addition to the demonstrable benefits accruing to clients, Chapter 10 also illustrates how contractors are benefiting from early appointments by obtaining the ability to influence project designs and programming and thereby to reduce their risks and secure increased profits. It will also consider the increasing importance of properly structured early contractor appointments in an economic downturn.

# CHAPTER TWO
# CONDITIONAL CONTRACTS AND EARLY PROJECT PROCESSES

## 2.1 The conditional preconstruction phase agreement

This chapter will consider the nature of the conditional preconstruction phase agreement in the context of contract theory and the potential areas of weakness of an agreement by reason of it being entered into at an early stage in the parties' relationship. The type of conditional preconstruction phase agreement of particular interest is one that is connected directly to the award of an unconditional construction phase building contract, such that the parties can rely on the expectation that once specific preconditions have been satisfied the construction phase can then proceed. This close link between the preconstruction phase agreement and the construction phase building contract is a means to establish a commercial justification for the contractor's contributions to preconstruction phase activities.

Although an early agreement for a contractor to provide preconstruction phase contributions can be freestanding, without any links to the construction phase contract, this approach breaks the continuity of the contractual system and requires two separate deals to be concluded. Separation of the preconstruction phase and construction phase contracts also creates the risk of a challenge under public procurement regulations for a public sector client if a third party can argue that the appointed contractor won a bid only to undertake the preconstruction phase works and that a new competition is required for the construction phase work[47].

A freestanding preconstruction phase agreement is also less likely to be commercially attractive to contractors. In particular, they may be concerned that, without direct links between the preconstruction phase and construction phase contracts, the benefits of their preconstruction contributions could be transferred to a competitor who might secure the construction phase work by undercutting their prices.

---

[47] See Chapter 9, Section 9.3.1 (Constitutional or regulatory constraints).

It is important that a conditional preconstruction phase agreement allows the team to move from the preconstruction phase with minimum negotiation. It is desirable that the contractual terms and conditions governing both the preconstruction phase and the construction phase are agreed at the outset, and that any revisions are proposed and accepted only as the legitimate outcome of joint risk management processes. If the team members are left free to negotiate more favourable terms in the run up to the construction phase, this will undermine the mutual confidence and stability that an early preconstruction phase agreement is intended to create.

It is, however, also important to recognise that a conditional preconstruction phase agreement is incomplete at the time when it is first entered into. If the parties adhere to its terms, but cannot finalise a mutually acceptable basis for proceeding to construction, then the preconstruction phase agreement must allow them appropriate leeway to withdraw. Arup in their report for OGC observed, in relation to PPC2000, that it:

> 'Is based on a two-stage tendering process whereby time and cost data is developed incrementally and reported on an open-book basis. This means that there can be a focus on value at all material points and the contract can still enable the parties to withdraw if the value profile is not satisfactory'[48].

A preconstruction phase agreement sets out to describe project planning processes and comes into effect at a point when the risks and responsibilities assumed by the parties are not fully known. It establishes the means and timescales whereby additional information is completed sufficient for the parties then to agree that the project should proceed to construction. The first question to consider is whether such a conditional document can be a binding, enforceable contract. If the contractual relationship is conditional and incomplete during the period until the construction phase building contract is concluded, might the preconstruction phase agreement be unenforceable for such conditionality or incompleteness[49]?

If a preconstruction phase agreement is only an agreement to negotiate the design or price or programme for the project, then it will be vulnerable for lack of certainty[50]. However, if a preconstruction phase

---

[48] Arup 2008, 37.
[49] Chitty states that an agreement can be unenforceable if it lacks meaning without agreement of further terms, but recognises that enforceable contractual machinery may be the means of achieving the required further agreement, Chitty (2008), 2–131, 216.
[50] Chitty cites *Courtney & Fairbairn Ltd* v. *Tolaini Bros (Hotels) Ltd* [1975] 1 W.L.R.297, 301 in support of the view that an agreement only to negotiate is not a contract 'because it is too uncertain to have any binding force', Chitty (2004) 2–136, 219.

agreement is incomplete as to design, price or programme, but contains the means to reach further agreement on particular points by utilising 'machinery laid down in the agreement' itself to arrive at more complete information, then it should not be challenged as unenforceable[51]. Chitty referred to 'arbitration' as an example of contractual machinery for reaching further agreement[52], but I suggest that more appropriate machinery in a preconstruction phase agreement would comprise a series of agreed activities (such as design development and subcontract tendering) allocated to specified parties within agreed timescales and required to meet agreed criteria (such as a project budget and client brief).

Nor is it essential that every detail can be determined using this machinery. In the activities required to prepare for a typical building project[53], some matters will need to be settled by further negotiation, but this should not undermine the effectiveness of the preconstruction phase agreement if the parties can be relied upon to apply a 'standard of reasonableness' or if the relevant matters are of 'subsidiary importance' such that they do not negate or overturn the intention of the contracting parties to be committed to the other terms that they have agreed[54]. This is an important qualification as it recognises the human element in project planning and the need for complex interactions between team members to arrive at increasingly complete information[55]. It also highlights the role of partnering as a project management system likely to encourage a culture of reasonable behaviour by the parties, both in applying agreed preconstruction phase contractual machinery and in negotiating subsidiary matters where required[56].

Not every activity can be programmed by contractual machinery, but that is not a reason to give up and abandon that machinery in favour of reliance only on negotiation and good faith. If primary objectives and criteria for meeting these objectives are agreed, with contractual machinery describing the agreed means to achieve those objectives, then the preconstruction phase agreement should remain contractually enforceable as to these matters even if other subsidiary matters require reasonable behaviour or negotiation in order to be resolved.

---

[51] Chitty (2008), 2–131, 216. Chitty also notes that an agreement dependent on such machinery is 'not, however, ineffective [as a contract] merely because such machinery fails to work', Chitty (2008), 2–131, 216.
[52] Chitty (2008), 2–131, 216.
[53] 'Close to a hundred different technologies' even in a simple building and 'more than a thousand work teams' in a complex project, Bennett & Pearce (2006), 4.
[54] See Chitty (2008), 2–129, 215, 216, as to the balance of binding primary obligations and negotiable secondary issues.
[55] See also Chapter 4, Section 4.2.2 (Integration with consultant designs).
[56] Partnering is considered in detail in Chapter 8, Section 8.3 (Preconstruction phase agreements and partnering).

Most parties will not adjust their behaviour to be more reasonable or compromise their negotiating position, even on secondary issues, without a good reason. Hence, the success of a preconstruction phase agreement will depend in part on it creating the commercial motivation for reasonable behaviour, for example in terms of the rewards payable during the construction phase of the project if it goes ahead or in terms of the potential for additional projects if the relationships are successfully preserved[57].

## 2.2  Recognised categories of contract

MacNeil classified contracts according to 'classical', 'neo-classical' and 'relational' categories of contract law[58], and this categorisation is in turn used by Williamson to describe the following models of contractual governance:

- Classical contract law 'which entails comprehensive contracting whereby all relevant future contingencies pertaining to the supply of a good or service are described and discounted with respect to both likelihood and futurity'[59];
- Neo-classical contract law, where 'not all future contingencies for which adaptations are required can be anticipated at the outset' and where 'the appropriate adaptations will not be evident for many contingencies until the circumstances materialise'[60];
- Relational contract law whereby (as in the case of neo-classical contract law) it is recognised that adaptations will be required to meet future contingencies, but where (unlike neo-classical contract law) the reference point is not the original agreement but 'the entire relation as it has developed … [through] time' which may or may not include an 'original agreement'[61].

None of the above categories fully describes the features of a conditional preconstruction phase agreement. It is not a classical contract as its purpose is to deal with the development of additional information to meet future contingencies (of design, risk, price and time), which the parties cannot foresee but in respect of which the contract should contain the means by which information will be built up. It is in part a neo-classical contract, requiring adaptation by agreed means to

---

[57] See Chapter 8, Section 8.3.3 as to the views of Bresnen and Marshall regarding the motivation for changing behaviour.
[58] MacNeil (1978), 854–905.
[59] Williamson (1979), 236.
[60] Williamson (1979), 237.
[61] Williamson (1979), 238.

capture new agreed information. It is also in part a relational contract as it seeks to govern not only agreed processes, but also development of the parties' relationship over time.

Cox & Thompson interpreted relational contract law as suggesting 'certain circumstances in business transactions when a contract is not necessarily required'[62]. However, I suggest that dependence on a personal relationship only, without the support of a clear written contract, is unlikely to be a sufficient basis for the commitments required to undertake a building project[63]. Milgrom & Roberts observed that under relational contracts:

> 'The parties do not agree on detailed plans of action but on goals and objectives, on general provisions that are broadly applicable, on the criteria to be used in deciding what to do when unforeseen contingencies arise, on who has what power to act and the bounds limiting the range of actions that can be taken, and on dispute resolution mechanisms to be used if disagreements do occur'[64].

This highlights the limits on treating a preconstruction phase agreement as a relational contract, as it does require the parties to agree a detailed plan of action as well as the criteria for particular actions, authority of individuals and means to resolve disagreements.

Contractual influence over the way the parties build up their relationship can include provisions that assist trust and mutual knowledge through the establishment of open-book cost information and the appraisal of other parties' risks[65]. The build up of such relationships can also be positively influenced by a clear regime of communications, meetings, delegated authority and early notification of issues that are potential problems[66].

Preconstruction phase agreements require the parties to recognise certain neo-classical features in their contracts, in particular gaps in what they need to agree that require the creation of a range of processes and techniques that enable the parties to complete missing information[67]. A degree of flexibility is required to steer a path between rigid contract terms on the one hand and omissions requiring and an 'agreement to agree' on the other hand. MacNeil envisaged the parties obtaining third party assistance to resolve disputes as to missing infor-

---

[62] Cox & Thompson (1998), 83.
[63] See Chapter 6, Section 6.4 (Non-binding arrangements).
[64] Milgrom & Roberts (1992), 13.
[65] See Chapter 4 Section 4.3 (Preconstruction pricing processes) and Section 4.4 (Preconstruction risk management processes).
[66] See Chapter 5, Section 5.3 (The role of communication systems).
[67] See MacNeil (1978), 865.

*Conditional Contracts and Early Project Processes*

mation[68], but this can be a cumbersome and potentially divisive approach as the third party needs to be briefed and paid, needs time to reach a conclusion and may not achieve a result that satisfies all parties. Instead, a preconstruction phase agreement can describe a series of processes and techniques to be undertaken by the parties themselves, and can recognise the conditionality of the commitments made until all parties are satisfied that their agreed processes and techniques have led them to create sufficient agreed information for the conditionality to be removed.

These processes will need to establish methods for dealing with matters outside the control of the contracting parties that may affect their interests, and the impact on the project if such matters are not satisfactorily resolved in the completed information necessary for the construction phase to proceed. Therefore, it is necessary to identify in a preconstruction phase agreement the parameters of any flexibility that the parties can allow in adjusting their expectations, and to establish a forum at which matters requiring flexibility can be considered by the parties with a view to reaching agreement. If the completed information exceeds the agreed parameters of flexibility or if such information cannot be completed, there also needs to be provision in a preconstruction phase agreement for abandoning the project and agreeing the consequent rights and entitlements of the parties.

For example, PPC2000 provides a right for the client to terminate the preconstruction phase agreement if any of the agreed preconditions are not satisfied for the construction phase to proceed or 'for any other reason not reasonably foreseeable by the Client', and restricts the other project team members' entitlements in this event to previously agreed amounts due in respect of preconstruction phase activities carried out prior to the date of termination[69].

## 2.3 *Effect of the number of parties*

Building projects generally involve numerous parties contributing to their design and construction. Performance of agreed obligations depends on the coordination of the different parties' interests and the

---

[68] For example, an architect or an arbitrator, MacNeil (1978), 865. MacNeil anticipated that under neo-classical contracts such a third party can assist in reconciling the differences created by opposing self-interested views. This is unlikely to be acceptable as the means of establishing the parties' unconditional commitments to the construction phase of a project. For example, it is likely there would be an immediate objection if such a third party recommended a price in excess of the client's budget or below the contractor's actual cost. However, see also Chapter 9, Section 9.6 (The role of the partnering adviser) regarding certain functions that a neutral third party can perform.

[69] PPC2000, clause 26.1 of the Partnering Terms.

reconciliation of their different motivations by means of contract terms. The extent to which coordination or motivation functions can be set out in a contract is an important question when considering the proposition that agreements can deal with preconstruction phase processes.

The issues were summarised by Milgrom & Roberts as follows:

> 'The coordination problem is to determine what things should be done, how they should be accomplished and who should do what ... The motivation problem is to ensure that the various individuals involved in these processes willingly do their parts in the whole undertaking, both reporting information accurately to allow the right plan to be devised and acting as they are supposed to act to carry out the plan'[70].

In creating contracts to deal with coordination and motivation, Milgrom & Roberts stated that 'These agreements may encompass the sort of actions each is to take, any payments that might flow from one to another, the rules and procedures they will use to decide matters in the future, and the behaviour that each might expect from the other'[71]. They recognised that contracts should govern not only actions and payments, but also rules and procedures for matters in the future and mutual expectations as to the parties' behaviour.

In creating a preconstruction phase agreement, the client will need to consider how best to integrate the preconstruction phase role of the main contractor, and the input from its subcontractors and suppliers, with roles of the client's design consultants and other consultants. The options are an integrated set of two-party contracts[72] or a single multi-party contract[73]. It has been suggested that the presence of additional parties itself creates circumstances that lead to the need for a contract dependent on relationships rather than clear written structures and processes[74]. It is, however, arguable that the presence of additional parties does not necessarily create a contract with more relational characteristics, if the planning structures and processes are clearly and comprehensively set out in that contract. There is not necessarily any direct link between the presence of additional parties and the chal-

---

[70] Milgrom & Roberts (1992), 126.
[71] Milgrom & Roberts (1992), 127.
[72] For example NEC3, Perform 21 and JCT CE, but not JCT 2005 which lacks a corresponding form of consultant appointment.
[73] For example PPC2000, the Perform 21 PSPCP Partnering Agreement and the JCT CE Project Team Agreement.
[74] MacNeil (1974), 792 as to the relational nature of multi-party arrangements. The integration of the roles of the different parties contributing to design is considered further in Chapter 4, Section 4.2 (Preconstruction design processes).

lenges of clear contractual planning processes, particularly where all parties are engaged on the same project with a view to meeting the same stated client objectives.

On the contrary, it is arguable that the additional work necessary to align a set of two-party contracts (and notify all team members of their contents) or to enter into a single multi-party contract has practical benefits as it ensures that all parties are aware of each other's roles and encourages each of them to check for errors, gaps and duplications. To demonstrate an even-handed approach by making it clear to all parties that their respective contract terms are consistent can also motivate the mutual trust among team members necessary for successful joint working[75].

## 2.4 The planning function of contracts

Arrighetti *et al.* saw the contract as 'a planning and incentive device' that can be used to link agreed objectives to the means of achieving them and the behaviour required from the parties'[76]. They went on to state that 'the law firstly creates a space within which the parties can plan the exchange, making due provision for future contingencies (the planning function), and secondly provides a set of sanctions aimed at inducing performance of the agreed obligations (the incentive function)'[77].

MacNeil also recognised the contract as having a planning function[78]: he distinguishes 'transactional' planning, which is binding and is allocative rather than mutual, from 'enterprise planning', which may be binding but some or all of which 'is characterised by some degree of tentativeness'[79]. It is interesting that MacNeil envisaged 'standardised construction contracts' as 'relational agreements containing a great deal of process planning'[80]. It is doubtful whether those who consider that building contracts should govern only the construction phase of a project would agree with this description.

---

[75] See Chapter 7, Section 7.6 (The impact of frameworks on changing behaviour) regarding the need for trust in order to motivate the sharing of sensitive information. See also Chapter 8, Section 8.3.9 (Confidentiality and disclosure) regarding the conflicting pressures of confidentiality and disclosure, which can only be resolved through clear contractual provisions describing what information is to be disclosed, on what terms and subject to what protections
[76] Arrighetti *et al.* (1997), 171.
[77] Arrighetti *et al.* (1997), 173.
[78] MacNeil identified the primal roots of contracts as 'reciprocity', 'role effectuation', 'limited freedom of exercise of choice' and 'effectuation of planning', MacNeil (1974), 809.
[79] MacNeil (1974), 739.
[80] MacNeil (1974), 760.

Preconstruction phase agreements should provide wherever possible for transactional planning, for example activities such as design deliverables which can be allocated in advance. However, they also need to recognise the need for enterprise planning, for example where planning for grant of design approvals (by the client or third parties) must remain tentative as achievement of the scheduled dates for such approvals cannot be predicted with complete certainty. In MacNeil's view 'lack of measurability makes binding planning difficult to accomplish in the first place and hard to carry out once done, and hence tends to make planning subject to change'[81]. However, planning under a preconstruction phase agreement can be measurable as such an agreement is focused on a specific goal, namely commencement of the project on site, and measurability is possible by reference to the likelihood of that goal being achieved.

The scope for successful mutual planning pursuant to a contract is affected by the need or not for negotiation, as negotiation gives rise to conflicting commercial interests and can bring project processes to a halt. MacNeil offers the following techniques for avoiding such conflicting interests, which have direct relevance to the operation of preconstruction phase agreements:

- Merging an allocative issue into enterprise planning as a result of which mutual, non-negotiating activities resolve the issue because the parties in pursuing such activities do not perceive the need for negotiation[82]. For example, while a project team will be well aware of scope for negotiation of prices, the use of subcontract tendering pursuant to a preconstruction phase agreement to build up open book cost information removes the need for such negotiation.
- Building up a business case for a particular course of action sufficient to demonstrate to all parties the benefits of that business case to the project as a whole, rather than leaving particular team members to haggle over prices or look for alternatives[83]. For example, a main contractor can build up a business case, pursuant to a preconstruction phase agreement, for use of a preferred subcontractor whose work it believes will benefit the project and be in the interests of the client and the design consultants. The subcontract price requires the support of such a business case in order to obtain client approval, and preparation of the business case also gives the main contractor the opportunity to demonstrate the qualitative benefits justifying its proposals.

---

[81] MacNeil (1974), 777.
[82] See MacNeil (1974), 780.
[83] See MacNeil (1974), 780.

*Conditional Contracts and Early Project Processes*

These techniques for overcoming the potential delay and uncertainty of negotiations require clear agreed processes to be set out in advance. A conditional two-stage agreement can more readily accommodate the required provisions as they can be described as processes to satisfy conditions rather than techniques to create a second contract. By contrast, a freestanding preconstruction phase agreement may leave the parties more at risk of protracted or unsuccessful negotiations because such negotiations are not part of an integrated set of contractual project processes, but are instead the means to finalise a construction phase building contract.

## 2.5  Choices and contractual conditionality

Relational contracts recognise that the contract reflects only the commencement of the relationship and will be followed by 'a complex succession of exercises of choice and agreement'[84]. These choices are required to accommodate and utilise increasing information. MacNeil recognises that none of these choices and agreements will encompass the entire contractual relationship until a 'final formal agreement' is established, and that the mutual planning processes that are the subject matter of the agreement must create a system whereby the exercise of choice is an incremental process. This allows the parties to gather increasing project information and to build up a full agreement stage by stage by an incremental process[85].

The above observations describe features of relational contract law that are reflected in the conditionality and agreed processes of preconstruction phase agreements. They suggest iterative processes of design development and procurement with which construction project teams will be familiar. It is, however, the uncertainty of a relational approach in building up full information that leads many clients and consultants to hold back from a contractual commitment to a contractor until an apparently complete agreement can be priced and concluded[86]. Where clear contractual processes are created to overcome concerns as to uncertainty, it is arguable that it is a combination of a relational contract and neo-classical contract, namely a 'process contract', that best describes an understanding whereby the contractor contributes to increased information in stages that satisfy stated conditions and allow

---

[84] MacNeil (1981), 1041.
[85] See MacNeil (1981), 1041 as to an incremental process of agreement using increasing amounts of information.
[86] As to questions regarding the apparent completeness of construction phase building contracts, see Chapter 3, Section 3.3 (Standard forms and the assumption of complete information).

the client to make a series of choices ultimately justifying an unconditional construction phase building contract.

A process contract can overcome the fear of opportunism that will otherwise dissuade the parties from entering into an incomplete agreement. It should also tackle head on the concern that an incomplete agreement will permit a lack of commitment to completing the remaining commercial details on competitive terms[87].

In considering the interaction between relational and neo-classical contract models, it is interesting that MacNeil chooses to cite as an example the construction of complex buildings. He perceives that there exists in such cases the practice 'of having architects, engineers, designers, etc., and the building contractor(s) all work together with the owner from the inception of site location, through building design, and through all other planning, which lasts until the final completion of the project'[88]. MacNeil's perception of such practices describes a team-based approach to a construction project that assumes early contractor and specialist appointments.

## 2.6 Limited efficiency caused by unknown items

To commence the contractual relationship of the client and contractor earlier than the typical construction phase building contract creates an opportunity to deal with unknown items by agreed processes, but increases the number and scale of those unknown items and the consequent incompleteness of the contract. The limited efficiency of incomplete agreements needs to be balanced against the inefficiency of postponing commencement of the client/contractor relationship until a larger number of items have apparently been resolved, such as issue of more detailed designs and agreement of more binding prices. Milgrom & Roberts noted in the context of 'bargaining costs' that 'It takes time and effort to imagine and list contingencies, to determine efficient courses of action, and to settle on divisions of costs and benefits', but that against this should be balanced 'the costs that are incurred in achieving commitment by noncontractual methods and the inefficiencies that result from attempts to protect against imperfect commitment'[89].

However, the apparently complete contract may be less efficient if unknowns still exist in the mind of the main contractor pricing the project and assessing its risk, or if apparently complete design information that has been issued for pricing is in fact incomplete or has been

---

[87] See Chapter 2, Section 2.7 (Risk and fear of opportunism).
[88] See MacNeil (1981), 1042.
[89] Milgrom & Roberts (1992), 147.

based on inaccurate assumptions made by the client or its consultants, particularly if it is then too late in the project process to make the necessary adaptations.

In assessing the weaknesses of tendering and contracting for the construction phase only, MacNeil draws the distinction between contracts that provide for the exercise of choice for economic reasons through 'bilateral power' and those that require the acceptance of obligations through 'unilateral power' by 'coercion' because of the adverse consequences of breach[90]. Main contractors bidding for a construction phase building contract, if they are required to price by reference to incomplete designs or inaccurate risk information, may make high risk or unprofitable (and thereby inefficient) tender commitments based on the adverse consequences of not winning the job. They then need to rely on making later claims against the client that exploit such incomplete designs and inaccurate risk information as the means by which to make a profit, and this creates further inefficiencies.

## 2.7 Risk and fear of opportunism

One challenge to creating a successful contract derives from what Williamson termed as 'bounded rationality'[91]. This describes the difficulties that people face by reason of limited foresight, imprecise language, the costs of calculating solutions and the costs of writing down a fully comprehensive contractual plan[92]. Wherever there is an omission or a lack of clarity in a contract, there is the risk of what Williamson describes as 'opportunism'[93]. In a preconstruction phase relationship where a full deal has not yet been concluded, there is a risk that the parties will fear opportunistic behaviour by each other that exploits any gaps or room for differing interpretations, and that they may not fully rely on each other as a consequence. This fear can have an adverse effect on the efficiency of project processes[94].

Milgrom & Roberts argued that with the best will in the world, the possibility of 'self-interested misbehaviour', which they also described as 'moral hazard' will remain and will oblige the parties to recognise that 'Real contracts are not perfect'[95]. They suggested that self-interested behaviour will restrict the scope for efficient joint planning and that what is required is the 'designing of systems that better align

---

[90] MacNeil (1981), 1052.
[91] Williamson (1985), 42.
[92] See also Chapter 9, Section 9.3.2 (Cost and time to create agreements) and Section 9.3.4 (Concerns as to conditionality).
[93] Williamson (1985), 42.
[94] See for example Milgrom & Roberts (1992), 128.
[95] Milgrom & Roberts (1992), 129.

individual interests, so that the constraints are looser and the available options richer'[96]. If the risk of opportunistic behaviour derives from omission or lack of clarity then I would argue this suggests the need for clear and complete preconstruction phase agreements rather than systems that involve looser constraints.

The risk of opportunism is greater where the parties are dependent on negotiation and greater still where the balance of commitment to the successful conclusion of that negotiation is uneven between them, for example where one party has invested more than the other in the expectation of the project going ahead or for other economic reasons is more reliant on a successful outcome[97]. A preconstruction phase agreement offers a means whereby the parties can create a contractual plan that does not leave any of them at a disadvantage in the event of opportunistic behaviour by another party.

First, if a preconstruction phase agreement is clear and binding, then it will be enforceable in the event of breach and will thereby be a disincentive to opportunistic behaviour[98]. Second, an agreement that describes systematic project planning processes reduces the need for negotiation and the risk of opportunism created by uncertain or imbalanced negotiating positions[99]. Third, such an agreement can encourage the parties' commitments if it provides for reasonable reward in respect of activities performed during the preconstruction phase[100], and if it states limits on the parties' respective rights in relation to abandonment of the construction phase in the event that the outputs from preconstruction phase activities do not satisfy agreed preconditions.

## 2.8 Conditional relationships without full consideration

In addition to regulating conditionality by agreeing early commitments, a further function of preconstruction phase agreements should be the harmonising of interests that would otherwise give way to the pursuit of what Williamson described as inconsistent and potentially 'antagonistic sub-goal pursuits'[101]. Williamson was commenting on what he called 'the economics of idiosyncrasy'[102]. A construction project

---

[96] Milgrom & Roberts (1992), 129.
[97] See Chapter 8, Section 8.3.4 (Challenges to successful partnering) regarding the roadblocks to partnering identified by N.J. Smith, which include uneven levels of commitment.
[98] The benefits of contractually binding preconstruction phase commitments are considered further in Chapter 6, Section 6.5 (Benefits of contractual clarity).
[99] See also Chapter 2, Section 2.4 (The planning function of contracts).
[100] See also Chapter 1, Section 1.4 (Early contractor appointments and payment).
[101] Williamson (1979), 239.
[102] Williamson (1979), 241.

can be considered an idiosyncratic supply of goods and services as it requires investments in the transaction of specific human and physical capital that depend on successful execution of the project before such investments are fully rewarded.

This links agreed roles and processes to the support provided by relationships between organisations and between individuals. The strength of these relationships is particularly important if and to the extent that investments of time and money are speculative. If one party is speculating more than the other, then there is the risk that such speculation will be abused by the other party in order to strike a better deal.

Selznick commented that contracts created to establish a 'pattern of cooperation for the achievement of common ends'[103] are unlikely to provide for full reciprocity, and indeed that to insist on this may defeat the purpose of such a contract. He stated that 'Reciprocity is never completely eliminated, but it tends to be overshadowed by dependency and rational coordination'[104]. The balance between reciprocity and speculation in the activities undertaken to achieve an unconditional construction phase appointment needs to be clearly stated in a preconstruction phase agreement.

It is possible, for example, that preconstruction phase commitments may not be rewarded with full consideration unless the project proceeds to the construction phase. A preconstruction phase agreement is then important as a medium through which to justify an apparent commercial imbalance. It can clarify the nature and extent of any speculative or partly rewarded work and can allow the parties to calculate and agree clearly the basis on which they commit to their respective investments in the transaction. If expressed as a conditional arrangement intended to lead into a later unconditional arrangement, it can set out the unconditional arrangement as a goal to assist the parties in recognising the benefits of working to satisfy the relevant conditions and to sustain the 'harmonising' of their different commercial interests throughout their speculative activities.

## 2.9 Need for long-term relations

Williamson points to neo-classical contract law and relational contracting as more appropriate to long-term contracts, on the basis that the increased costs of setting up a complex governance structure should only be reserved for 'complex relations', which he defines by reference to the following attributes:

---

[103] Selznick (1969), 58.
[104] Selznick (1969), 58.

*Early Contractor Involvement in Building Procurement*

(1) Uncertainty as to the outcome of the relationship;
(2) The need for the parties to invest in the transactions with a view to ensuring their success;
(3) The number of projects giving rise to frequently recurring transactions.[105]

It is accepted that there is uncertainty of outcome in a preconstruction phase agreement and the need for the prospect of significant work to justify early investment in the transaction. However, it is not necessarily the case that frequency of recurrence (i.e. more than one project) is essential to justify a preconstruction phase agreement. The transaction-specific investment may be justifiable for a single project if that investment is proportionate to the size and significance of the project and particularly if it can be scoped and understood (and its risks thereby potentially reduced) by means of a clear agreement.

It is attractive economically to the parties if their initial investment in a system of project planning, established under a preconstruction phase agreement, can be repeated on successive projects. It is also encouraging psychologically for the parties to be able to keep a team together on additional projects that has worked efficiently on one or more earlier projects, although the mobility of people between organisations and their natural career progression often mean that the apparent continuity and established learning of such a team are not always entirely what they seem.

It is, however, suggested that the factors that influence the parties in making a commitment to project planning are more complex than the prospect of repeat business. They can also include issues such as the size of the project and its importance to the parties, for example in terms of prestige and opportunities to enter new markets or demonstrate new technologies[106].

## 2.10  Alignment of different interests

Neo-classical contract law and relational contracting describe a contract that responds to adaptations, and partnering enthusiasts often suggest that the team members will work together outside any written rules to agree the right way forward[107]. It is, however, questionable whether such a contract can rest on the assumption that the parties will

---

[105] Williamson (1979), 239.
[106] As illustrated by Willmott Dixon's commitment to two-stage procurement of partnered schools projects, Project case study 2 (Contribution of preconstruction phase agreement), Appendix A.
[107] See also Chapter 6, Section 6.4.2 (Unwritten understandings).

adapt in a consistent manner. Many project situations will indeed depend on coordinated responses, but it is possible that coordination will not be successful if different parties react differently in the way that they assess what is expected of them. Individuals can read and react to signals differently, particularly verbal signals in a meeting or conversation, even in circumstances where the declared purpose of the parties is to achieve 'a timely and compatible combined response'[108].

This highlights that early arrangements, particularly those based on incomplete and changing information, may be at risk of misunderstandings and consequent problems or disputes caused merely by differing interpretations, rather than by bad faith or incompetence. Bennett envisaged project procedures and standards that 'establish design details, predetermined roles, patterns of meetings, flows of information and planning and control systems' and support 'an almost automatic way of working'[109]. However, this appears to be an over-optimistic view of the way that most projects are managed and implemented, and it will be difficult to rely on these automatic ways of working if they are not clearly rehearsed in writing. By contrast, Williamson envisaged the need to reduce time spent bargaining from positions of self-interest by means of 'The conscious, deliberate and purposeful efforts to craft adaptive internal coordinating mechanisms'[110].

In the absence of project planning processes so well established that they are automatically understood consistently by all parties and can be implemented without the risk of different interpretations, there is a role for a contract that describes those processes and provides a medium to support consistent interpretations and minimise the scope for misunderstandings.

It is necessary that clearly agreed systems are put in place in order to achieve what Williamson calls 'cooperative adaptation' whereby parties can coordinate the alignment of their contractual positions that is necessary to fill gaps in project information[111]. While it may remain in their collective interest for the different parties to move from incomplete to complete information, correcting any errors or misunderstandings that are necessary to achieve efficient realignment of their positions, they may also perceive the gains arising from making any such adjustment differently[112].

Parties cannot be obligated to compromise their rights through cooperative adaptation, but the process of alignment can be assisted by structured meetings of the parties' representatives to address the issues

---

[108] See Williamson (1993), 48.
[109] Bennett (2000), 106.
[110] Williamson (1993), 48.
[111] See Williamson (1993), 48.
[112] See Williamson (1993), 48.

when they arise[113]. The clearer the guidance by way of procedures and terms of reference for any such meetings, linked to mechanisms for the incremental agreement of new information, the greater the chances of a solution that will serve the collective interests of the parties in preserving the parties' relationships while also respecting their individual commercial interests. All of these can be set out in a preconstruction phase agreement[114].

## 2.11 Preconstruction phase agreements as project management tools

A preconstruction phase agreement can act as a project management tool that guides the parties through the performance and integration of certain agreed activities. R.J. Smith recommended that 'A good contract should be in essence a handbook for performance'[115]. Cox & Thompson observed that 'One slightly less common function of contracts is that of a project management tool. Contracts ... are rarely used as the mechanism which prescribes the parties' actions in a programmed manner'[116]. Cox & Thompson recognised the potential for certain standard form building contracts such as NEC3 to act as a 'project management aide memoire' which can be 'based on the same principles as a flow chart and guide the parties as to 'what to do next''[117]. N.J. Smith proposed that 'Contract terms should be designed to motivate all parties to try to achieve the objectives of the project and to provide a basis for project management'[118].

All of these recommendations recognise that project management processes can be expressed in contractual provisions. None of them specifically recognise the value of the contract as a 'handbook', 'project management tool' and 'basis for project management' during the preconstruction phase, yet this is the period during which a series of critical processes require careful project management[119].

Commons subdivides transactions as 'bargaining' (a contract of a voluntary nature), 'rationing' (a contract exercising authority) and

---

[113] The benefits of organising structured meetings of a group of named individuals known as a 'core group' are considered further in Chapter 5, Section 5.3.3 (Creation of a contractual core group).
[114] Procedures for meetings and related communication mechanisms are considered further in Chapter 5, Section 5.3 (The role of communication systems).
[115] Smith R.J. (1995), 42.
[116] Cox & Thompson (1998), 45.
[117] Cox & Thompson (1998), 45, 46.
[118] Smith N.J. (2002), 14.
[119] The links between preconstruction phase agreements and project management are considered in detail in Chapter 8. Section 8.2 (Preconstruction phase agreements and project management).

## Conditional Contracts and Early Project Processes

'managerial' (a contract for coordination)[120]. Standard form building contracts have traditionally sought to strike as complete as possible a 'bargain', and then to 'ration' subsequent activities with a view to controlling cost, time and quality. In a preconstruction phase agreement, the 'bargain' will be incomplete, and exercise of authority through 'rationing' is therefore not likely to be a sufficient mechanism alone to govern the activities necessary to complete the bargain in time for the construction phase. It is submitted that process contracts in the form of preconstruction phase agreements govern primarily 'managerial' transactions.

## 2.12 Contracts and new procurement systems

The hybrid nature of conditional preconstruction phase agreements should not be an obstacle to their effectiveness or to their validity as contracts, but instead is evidence of evolution of contracts generally and of building contracts in particular. If the introduction of additional preconstruction phase agreements can be seen as a logical evolutionary progression, then its potential benefits need to be examined and its characteristics clarified and made available as widely as possible.

It has been argued that the treatment of a building contract as a simple and complete transaction has put enormous pressure on the ability of that contract to cope with situations for which it was not designed[121]. Wider recognition of the potential of a new type of building contract would relieve this pressure.

Commentators have suggested that contracts should only support commercial objectives and procurement models that are already well established, and that it is not the function of a contract to try to create a model that is in advance of the established position of the construction industry[122].

This argument is vulnerable as it does not explain how to pinpoint the current established commercial objectives of the industry at any given time, or for how long a commercial position needs to have been established before it is appropriate to reflect it in a building contract. It is therefore arguable that a conditional preconstruction phase agreement that sets out the processes governing two-stage procurement should be available to clients, main contractors, consultants, specialist subcontractors and suppliers, irrespective of whether this is an established industry approach, so as to enable the parties to make informed

---

[120] Commons (1934), 34.
[121] See MacNeil (1974), 815 as to the need for 'frank recognition' of the relational characteristics of contractual behaviour.
[122] See for example Jones *et al.* (2003), 188.

decisions as to if and how they may wish to adopt it. The Construction Industry Council took this view in promoting the creation of a multi-party project partnering contract, and stated that 'Creating a contract that can accommodate [new] aspirations is clearly of paramount importance in the development of partnering'[123].

This chapter has established that a conditional preconstruction phase agreement can be an enforceable contract governing contractor contributions to the preconstruction phase processes necessary to prepare a project for its construction phase. While vulnerable to exploitation of any matters still subject to negotiation, such an agreement can offer techniques to avoid such exploitation and can provide additional clarity that itself helps to harmonise differing commercial interests.

---

[123] CIC (2002), 12.

# CHAPTER THREE
# PROBLEMS AND DISPUTES UNDER CONSTRUCTION PHASE BUILDING CONTRACTS

## 3.1 Introduction

Having considered the characteristics of a conditional preconstruction phase agreement in Chapter 2, this chapter will contrast such characteristics with those of construction phase standard form building contracts. It will look at the origins of standard forms, and their shortcomings in the eyes of clients and commentators. It will be argued that there are links between the limitations of construction phase standard forms and the predominant causes of construction disputes, insofar as many of these disputes are directly attributable to the absence of clearly agreed preparatory processes involving the main contractor and its design subcontractors.

## 3.2 The role of standard form building contracts

The widespread use of published standard form building contracts, as distinct from bespoke building contracts, is a way of significantly reducing transaction costs by allowing the parties to base their contractual relationship on knowledge already acquired in relation to previous transactions using the same published form. Clients may be advised to adhere to published standard form building contracts due to:

- The prohibitive cost of creating a bespoke form of contract for each new project;
- The concern that main contractors will be unwilling to bid against or enter into an unfamiliar form or will add a premium to their prices for doing so;
- The limited time available for creating building contracts when there is pressure to obtain bids and to start work on site;
- The high perceived risks in amending standard forms, other than in relation to well-rehearsed points of negotiation, on the basis that an amendment in one clause of a building contract is likely to impact on numerous other clauses.

The publication of standard form building contracts, in particular the JCT forms now available for over 75 years, has enabled large numbers of contractual relationships to be concluded in a short space of time and with minimal cost involved in drafting or negotiation, but are they fulfilling their intended purposes if, for reasons of cost or expediency, they are not actually read by their signatories? It is suggested that they are not and that in part this is due to building contracts only being entered into late in the project processes at (or around) the time of commencement of work on site. R.J. Smith observed that 'Even though contracts serve multiple useful functions, their content and development is sometimes given insufficient attention'[124].

In some cases the position is made worse by the use of complex wording. Duncan-Wallace referred to certain building contract provisions as creating 'a private world of mystery where only the professionally qualified may tread'[125]. If relatively few individuals read the detail of standard form building contracts, there is a risk to project teams in the acceptance and use of established contract provisions that are not fully understood by those committing to them.

## 3.3 Standard forms and the assumption of complete information

Barlow *et al.* observed that 'Traditional models view the construction process as the purchase of a product, governed by legal contracts' with minimal uncertainty as to what product is required. They noted that such contracts seek to include provisions whereby 'any uncertainty in the means by which it is implemented is passed onto contractors or subcontractors as risk'[126].

Many standard form building contracts include provisions intended to fix the scope of the work, time and price, and at first sight appear to be complete contracts, anticipating and dealing with all eventualities[127]. They have as their starting point an advanced state of design, suggesting that the client's requirements and the contractor's proposals for meeting those requirements have been fully defined. It is therefore tempting to categorise construction phase building contracts as having only the features of classical contract law, and as describing a comprehensive arrangement between the parties.

However, standard form building contracts are not entirely complete, as they include provisions to govern change to deal with the

---

[124] Smith R.J. (1995), 42.
[125] Duncan-Wallace (1996), Vol. 2, 505.
[126] Barlow *et al.* (1997), 5.
[127] For example, JCT 2005, NEC3 and GC/Works/1.

consequences of a variety of risk events and to complete by means of later pricing and instruction certain parts of the project that are undefined or 'provisional' at commencement of the construction phase[128]. By seeking to accommodate contractual systems designed to deal with the unforeseeable consequences of change and risk events and provisional items, the parties acknowledge that not all future contingencies can be fully anticipated, and to this extent their contract has the characteristics of neo-classical contract law. For example, if the parties cannot agree the scope or cost of a provisional item at the commencement of the construction phase, their contract remains in this respect conditional until such matter is agreed. If building contracts governing the construction phase are not entirely complete and unconditional, there is less difference in principle between such contracts and agreements governing the processes required to implement the preconstruction phase of a project.

## 3.4 Origins of standard forms and lack of trust

The creation and negotiation of many building contracts is based on the premise that the other party is the opposition and cannot be trusted, and that consequently there is little or no room for allowing flexibility or achieving harmonised interests[129]. It is arguable that such building contracts are self-limiting where they only seek to coerce performance through allocation of liability irrespective of the ability to manage risk, rather than to facilitate performance through recognition of unknown matters and establishment of agreed processes by which they can be dealt with. An example of this is given by Smith *et al.*, who stated that 'This allocation of risk is defined in many of the standard forms of contract' and went on to state that 'the decision regarding the form of contract is a crucial decision in the management of risk'[130]. In fact, adopting the fixed risk allocation that forms the basis of most standard forms is a risk management decision only in the negative sense that it usually denies the contractor and its subcontractors any role in the client's risk management process[131].

When combined with a 'take it or leave it' single-stage tender procedure, such contracts do not assist the creation of an integrated and fully functional team. MacNeil expressed concerns that to obtain 'short

---

[128] For example, JCT 2005 SBC/Q clauses 2.29, 4.24 and 5 and NEC3 core clause 60.
[129] For example, N.J. Smith observes that 'generally the interests of the promoter and the contractor tend to be opposed to each other', Smith N.J. (2002), 176.
[130] Smith *et al.* (2006), 18.
[131] See also Chapter 2 Section 2.6 (Limited efficiency caused by unknown items) as to the impact of coercion on efficiency.

and sharp consent to the unilaterally developed terms' in a single-stage building contract is 'a process heavily laden with conflict'[132].

Lack of trust is in fact not a logical basis for denying the potential for early contractor involvement. If contractors will exploit single-stage tendering by winning work at lowest cost and then seeking changes and claims for delay and disruption, then their approach will not be changed simply by loading more onerous obligations into the same construction phase building contracts. A cynical, but compelling argument for early contractor involvement is that it offers the means to acquire additional information from the contractor that will make it difficult for changes and delays to be exploited rather than honestly assessed[133].

What are the origins of standard form building contracts based on a lack of trust? The JCT forms stem from social and economic changes in the nineteenth century, shifting power away from the client, and increasing pressure for architects and main contractors to establish a mutually acceptable standard form. Despite a 'decidedly cool'[134] relationship between the architects' and contractors' representative bodies, such a standard form was produced in 1931 and led to the formation of the JCT[135]. It is arguable that the efforts required to achieve consensus between its various members have not always helped the JCT see the role of its contracts as governing mutually beneficial project processes rather than reconciling divergent commercial interests, and have often impeded keeping the JCT forms up to date with the development of new procurement models and with industry perception of modern best practice.

Bennett observed that the operation of drafting committees is 'a distinctive feature of UK construction practice'[136] in relation to standard form building contracts, particularly the involvement of unpaid individuals representing various sponsoring industry bodies. Bennett cites frequent meetings, the slow pace of committee work and the separate reviews by constituent bodies as giving rise to 'a conservative view of current good practice'. He recognises the strength of such contracts as 'safe to use' and representative of a 'middle-of-the-road majority view', but expresses the opinion that they 'do not represent the best'[137].

---

[132] MacNeil (1974), 777.
[133] See Chapter 9, Section 9.4.3 (Industry conservatism) and the quoted chief QS of a leading construction company regarding exploitation of changes and delays to improve the commercial return on a single-stage lowest price tender.
[134] Nisbet (1993), 60.
[135] The JCT was joined by the quantity surveyors of the RICS in 1947, by representatives of local authority associations and specialist subcontractors in 1963 and eventually by the Association of Consulting Engineers in 1975/76.
[136] Bennett (2000), 178.
[137] Bennett (2000), 178.

*Causes of Problems and Disputes*

The JCT approach to standard form contracts has changed significantly in recent years, with new options emerging through the introduction of the JCT Major Project Construction Contract in 2005, the JCT Constructing Excellence Contract in 2006 and the JCT Pre-Construction Services Agreement in 2008[138]. However, a significant and surprising limitation on the JCT 2005 suite of contracts remained, namely that after so many years it still did not include a form of consultant appointment[139]. If and to the extent that JCT contracts do not include forms of appointment for consultants, they cannot integrate the roles and responsibilities of all team members[140].

## 3.5   Criticisms of effectiveness of standard forms

In 1994, Latham recommended that a modern construction contract should include 'A wholly interrelated package of documents which clearly defines the roles and duties of all involved, and which is suitable for all types of project and for any procurement route'[141]. He pointed to the New Engineering Contract as a suitable model, and recommended that reform was required in the JCT in order to facilitate creation of a new 'complete family of standard documents' which 'should include a total matrix of interlocking consultants' agreements and contracts, including subcontracts, available for all kinds of building work'[142]. The same year Rhys-Jones suggested that there is 'a need to challenge long-held assumptions within the construction industry and the legal profession on issues such as the effectiveness of standard forms of contract, acceptable language, preferred methods of dispute resolution, styles of drafting and management'[143].

In 1996, Duncan-Wallace argued forcibly for the radical reform of the JCT and ICE forms, with particular concern for their perceived bias in favour of the contractor and for their encouragement of a single-stage, lowest price procurement strategy that meant selection on the basis of low prices was frequently followed by contractual claims. He commented that 'the attraction of the lowest possible tender (contract) price accompanied by the maximum number of expressly permitted

---

[138] JCT 2005 MPCC, JCT CE and JCT PSCA. As to the latter, see also Chapter 6, Section 6.2 (Building contract options) and Appendix B.

[139] JCT CE can be used for the appointment of a consultant as well as a contractor, but there are no consultant appointments designed to be used with any of the JCT 2005 suite of contracts.

[140] Announcement of a JCT 2009 public sector consultant appointment was the first step towards remedying this significant omission from the JCT suite of contracts.

[141] Latham (1994), 37, Section 5.18.3.

[142] Latham (1994), Sections 5.26. and 5.28.

[143] Rhys-Jones (1994), 10.

post-contract increases of price (claims) has become regarded as axiomatic in those industry circles concerned to advance the use of the standard forms generally'[144]. It is interesting that Duncan-Wallace linked his criticism of standard forms to his concerns regarding the effect of single-stage procurement.

In 2000, Bennett expressed the view that standard forms of contract were too focused on creating excuses for failure in performance and for the acceptance of late project completion, cost overruns and defects in the completed project. In Bennett's view, such attitudes expressed in contracts are not consistent with the modern construction industry and should not be acceptable to clients if they create any obligation to accept the consequences of substandard work[145]. The suggestion that the inadequacies of standard form building contracts can adversely affect the interests of the client justifies further examination of the links between such contract forms and the causes of client dissatisfaction.

Latham identified various sources of dissatisfaction on the part of construction clients, including poor value, poor quality and late completion. Latham's 1994 report included a tabular comparison of client expectations, such as 'Value-for-money', 'Pleasing to look at', '(largely) Free from faults' and 'Timely Delivery'[146]. He compared levels of satisfaction with the modern motor car to levels of satisfaction with domestic, commercial and industrial modern buildings, and found that the motor car scored consistently much higher on all fronts.

Egan commented in *Rethinking Construction* that growing dissatisfaction of both public and private sector construction clients derived from the fact that 'Projects are widely seen as unpredictable in terms of delivery on time, within budget and to the standards of quality expected' and that 'Clients believe that significant value improvement and cost reduction can be gained by the integration of design and construction'[147]. However, neither commentator established links between these problems and the forms of contract used on the relevant projects.

## 3.6   Causes of claims

Duncan-Wallace referred to 'the reputation for aggressive "claimsmanship" of United Kingdom contractors generally'[148]. It is often assumed by such commentators that project cost increases, and in particular contractor claims for additional time and money, are evidence of

---

[144] Duncan-Wallace (1996) Vol. 2, 509.
[145] See Bennett (2000), 174.
[146] Latham (1994), 12.
[147] Egan (1998), Sections 5 and 7, 10–11.
[148] Duncan-Wallace (1996) Vol. 2, 506.

opportunism, bad faith and incompetence on the part of the main contractor. While commercial self-interest may in some circumstances encourage a contractor to apply the first two and disguise the third, it is rarely straightforward or profitable for contractors to make their money through claims and the underlying picture is more complex. In R.J. Smith's view, if risk events do occur and disputes and delays become likely, then the 'cost of victory' is considerable[149].

As to the main reasons for unanticipated increases in the construction cost of projects, the 1995 study *Construction Procurement by Government* identified the following:

- 'The objectives were unrealistic or changed during the course of the project;
- Estimates for project approval were too optimistic;
- The project brief was incomplete, unclear or inconsistent;
- The design did not meet planning or statutory requirements;
- The design was incomplete at the time of tender;
- The design lacked coordination, buildability or maintainability;
- Risk allocation was ambiguous; and
- Management control was inadequate'[150].

All but the last of the listed reasons relate to preconstruction phase activities and nearly all of them would be likely to be examined more closely if there was main contractor participation at an earlier point in the preconstruction phase.

For example, the robustness of the client's objectives and estimates and of the successful main contractor's tender prices (any of which may be over-optimistic and/or poorly researched) can be tested through main contractor involvement in systematic examination of actual underlying costs in advance of start on site, allowing time for value engineering to bring cost overruns back within budget. Similarly, omissions or inconsistencies in the project brief, and incompleteness or lack of coordination, buildability or maintainability in the designs (and their non-compliance with planning requirements), can be challenged by the main contractor and rectified by the client and consultants working in conjunction with the main contractor as preconditions to the project proceeding on site. Earlier appointment of a main contractor could also permit joint examination of risks, the clarification of how cost has been allocated to perceived risks and agreement of the actions that need to be taken to reduce this cost. Such analysis in turn could reduce the ability of either party to ignore or exploit any ambiguities in risk allocation during the construction phase of the project.

---

[149] Smith R.J. (1995), 44.
[150] Efficiency Unit (1995), Section 145, 53.

Meanwhile, as to the causes of main contractor claims, Kumaraswamy identified in 1997 the following 'Top ten' causes as perceived by contractors, clients and consultants – listed in descending order of overall perceived significance:

(i) 'inaccurate design information
(ii) inadequate design information
(iii) inadequate site investigations
(iv) slow client response (decisions)
(v) poor communications
(vi) unrealistic time targets
(vii) inadequate contract administration
(viii) uncontrollable external events
(ix) Incomplete tender information
(x) unclear risk allocation'[151].

The claims identified by Kumaraswamy have given rise to disputes which need to be resolved by reference to a third party, whether an adjudicator or an arbitrator or a court. Such disputes inevitably damage working relationships, distract the parties from their work on the project and involve significant legal and other costs. All of the listed causes of claims, if those claims are successful, give rise to project delays and client liability for additional payments. It is therefore worth considering whether there are means by which the claims themselves could be averted, or their effects better understood and better mitigated, by the actions of the parties themselves.

## 3.7 Links between claims and preconstruction phase activities

The great majority of Kumaraswamy's listed causes of claims (and certainly all of the top six) are directly linked to activities undertaken during the preconstruction phase of the project, namely prior to the creation of most standard form building contracts and (in most single-stage models) prior to any engagement between the client and main contractor. Specifically:

(i) Inaccurate design information – without a period prior to the construction phase for the main contractor (and any subcontractors and suppliers contributing to design) to review and comment on and contribute to designs, inaccuracies are less likely to be identified.

---

[151] Kumaraswamy (1997), 5.

*Causes of Problems and Disputes*

  (ii) Inadequate design information – without main contractor and subcontractor design contributions prior to commencement of the construction phase, inadequacies are less likely to be revealed. In addition, where further detailed design information is required during the construction phase, the absence of previous main contractor and subcontractor engagement increases the risk of misunderstandings between design consultants and the main contractor and subcontractors as to what information is required from whom and as to when it becomes time critical[152].

 (iii) Inadequate site investigations – this is a key component of risk assessment and allocation in respect of which the client and its consultants may take a different view from the main contractor as to the nature and detail of the investigations that are appropriate. A procurement system that provides for joint working by the client, the consultants and the main contractor during the preconstruction phase, leading to agreement of what site investigations are necessary and how to deal with their results, avoids the need for the main contractor simply to price the client's investigations and argue about inadequacies later. Instead, the parties can use the preconstruction phase to agree who should commission any necessary additional site investigations so as to reduce risk, risk pricing and the likelihood of claims deriving from inadequate risk pricing[153].

 (iv) Slow client response (decisions) – the importance of timely client decisions can be assessed, tested and agreed in a thorough manner only if there is a period during the preconstruction phase when the client and main contractor can work together to establish and agree the key dates and periods for critical client decisions, and if these are then set out in a suitable document forming part of the construction phase building contract[154].

  (v) Poor communications – many standard form building contracts do not provide for a detailed communications strategy and assume this is something for the team to establish outside the contractual provisions. The absence of a preconstruction phase period of engagement among all team members (including the main contractor), during which key individuals can be identified and form working relationships, means that the construction

---

[152] Integration of design activities is considered further in Chapter 4, Section 4.2 (Preconstruction design processes).

[153] Management risk in respect of site conditions is considered further in Chapter 4, Section 4.4 (Preconstruction risk management processes).

[154] An example of such a document is the JCT Information Release Schedule, JCT 2005, SBC/Q clause 2.11. See also Chapter 5, Section 5.2.3 (Client involvement in preconstruction phase processes) regarding the importance of the client working with the contractor and design consultants so as to avoid 'the proliferation of claims', Smith *et al.* (2006), 136.

phase may commence without channels and forums for communication being properly established[155].

(vi) Unrealistic time targets – if the client and the main contractor have no opportunity for joint programming during the preconstruction phase, deadlines are often established by a unilateral client decision to reflect its particular demands and/or its consultants' advice as to what is achievable. These may be confirmed or adjusted by the successful contractor in a competitive bid, without the contractor having knowledge of the client's resources and expertise, and those of its consultants, which could impact on required deadlines dependent on the achievement of, for example, prompt client decisions, detailed consultant design outputs and other matters affecting the project or the site. To exacerbate the scope for disagreements and misunderstandings, the majority of standard form building contracts allow for a construction programme to be agreed only after commencement of the project on site[156].

(vii) Inadequate contract administration – although this is a feature of the construction phase of the project, it is arguable that earlier client engagement of the main contractor would enable the client to review the need for contract administration resources and to establish what level would be appropriate according to, for example, the quality control systems proposed by the main contractor[157].

(viii) Uncontrollable external events – the complex nature and physical risks of most building projects leave them vulnerable to external events and consequent claims for additional time and money. However, such claims are less likely to be disputed if the proposed allocation of risk in respect of external events has been tested during the preconstruction phase to establish whether such risks can be reduced and if the financial allowance made for such risks can be calculated in the least controversial manner[158].

(ix) Incomplete tender information – in a single-stage competitive tender process, bidding main contractors and their subcontractors have little opportunity to question the completeness of tender information. A two-stage process allows the client to issue tender information appropriate for early main contractor selec-

---

[155] The features of such a communication strategy are considered further in Chapter 5, Section 5.3 (The role of communication systems).
[156] For example, JCT 2005, SBC/Q clause 2.9 and NEC3, core clause 31.1. Joint programming is considered further in Chapter 5, Section 5.4 (The role of binding programmes).
[157] The benefits of closer client involvement in a project are considered further in Chapter 5, Section 5.2 (The role of the client).
[158] The scope for joint risk management is considered further in Chapter 4, Section 4.4 (Preconstruction risk management processes).

tion and then to work with the selected main contractor in preparing and issuing further tender information for second-stage subcontractor selection. A two-stage tender approach thereby enables the client to seek an unambiguous commitment from the main contractor and its subcontractors that such tender information is sufficiently complete for the project to proceed on site, thus substantially reducing the risk of later claims that such information was inadequate. The PPC2000 Commencement Agreement contains such an undertaking, whereby 'The Partnering Team members ... agree ... that ... to the best of their knowledge the Project is ready to commence on Site'[159].

(x) Unclear risk allocation – most standard form building contracts are clear in their allocation of certain risks during the construction phase, for example as to grounds for the contractor to claim extensions of time for project completion and additional money[160]. However, as in the case of site investigations (item (iii) above), a preconstruction phase joint risk management process can be used to clarify any points of doubt as to risk allocation, for example to resolve any gaps or duplications in team members' roles and responsibilities.

It is therefore arguable that all the top ten causes of construction claims as identified by Kumaraswamy's contractors, clients and consultants would to some degree be less likely to arise if the client and main contractor (with consultants, specialist subcontractors and suppliers involved as appropriate) entered into early mutual commitments to ensure that the following preconstruction phase activities occurred:

- Joint design development (claims (i) and (ii));
- Two-stage tendering (claim (ix));
- Joint risk management (claims (iii), (viii) and (x));
- Advance agreement of a construction phase programme (claims (iv) and (vi));
- Development and implementation of a communications strategy (claim (v));
- Closer client involvement with its project team (claim (vii)).

## 3.8 Links between claims and building contracts

There is set out in Figure 3.1 at the end of this chapter the preconstruction phase and construction phase procurement activities that precede

---

[159] PPC2000, Appendix 3, Part 2, Form of Commencement Agreement.
[160] See, for example the specific grounds for claiming additional time and money listed in JCT SBC/Q 2005, clauses 2.29 and 4.24; NEC3 core clause 60; and PPC2000 Partnering Terms, clause 18.

and follow a typical single-stage construction phase contractor appointment, noting the risks that arise if contractors do not have an involvement in the earlier activities. These risks include:

- Designs – these are developed in detail by the design consultants (including obtaining third party consents), without being tested with main contractors or specialist subcontractors for their buildability or affordability.
- Costs – the cost plan is developed by cost consultants on the basis of increasingly detailed designs, without being tested with main contractors for its accuracy until main contract tenders are returned immediately prior to commencement of the construction phase.
- Risks – risks are assessed by the client and the consultants, making assumptions as to the likely interpretation of such risk assessments by the main contractor and its subcontractors.
- Main contract tender – the invitation to tender requires main contractor bidders within a limited time to assimilate complex project information, to obtain a full set of priced subcontract bids and to propose a complete and binding project price.
- Subcontract tenders – main contractor bidders need to require their respective subcontractor bidders to assimilate complex project information within an even more limited time and with a proportionally remote prospect of success, and to propose complete and binding subcontract prices.
- Joint activities – any post-tender reviews by the selected contractor of the client's and consultants' design and risk assumptions will delay start on site, allowing very limited scope for any joint client, consultant and contractor activities, such as value engineering or risk management to reduce excessive costs or to resolve consultant design errors revealed in main contract tenders. Such reviews also become confused with commercial negotiation of the contractor's price.
- Programme – the construction phase programme is not agreed until after start on site, creating uncertainty as to key dates required for activities such as the release of further consultant or contractor design details and for the pricing and approval of provisional sum items.
- Subcontractor appointments – subcontractor appointments are likely to be finalised only after start on site, allowing the potential for re-pricing and change of subcontractors (with no benefit to the client) if the main contractor seeks to increase its profit by finding cheaper subcontract deals or if subcontractors withdraw before they are signed up.

The links between construction claims and building contracts have been identified in successive industry reports. NEDC in 1991 observed that 'The adversarial relationship established by the traditional contractual framework does not stop with the completion of the project. Claims and counter-claims continue often for years afterwards, exhausting the industry from both energy, resource and cost aspects'[161].

Judge Anthony Thornton, speaking by reference to Professor John Uff's 2003 Brown Memorial Lecture, recognised Uff's concerns that excessive construction costs in the UK are 'further increased as a result of the large number of claims and disputes that construction generates', and noted in particular complaints originating from:

- 'inefficient working practices'
- 'poor planning'
- 'inadequate identification of the scope of the required work'
- 'unsatisfactory designs, detailing and specifications' and
- 'low standards of workmanship'[162]

He observed that these complaints are compounded by lack of trust and cooperation and lack of 'information sharing between participants at every level of the design, planning and construction chain'[163]. It is significant that Uff's concerns, as represented by Thornton, focus on problems in the early planning stages of the project and on the deficiencies of communication between project participants. Uff's proposed solution was a 'code of ethical conduct' to govern the roles of all construction professionals appointed under all building contracts. However, neither Uff nor Thornton acknowledged that, even with a consultant code of conduct, the same problems may still arise by reason of the absence of the contractor during the preconstruction phase.

Unforeseen problems are likely to arise during construction projects, and it is suggested that standard form building contracts governing only the construction phase of a project do not always offer the available solutions. As a significant number of problems can be attributed to preconstruction phase activities and the absence of the main contractor from those activities, construction phase building contracts can offer no solution at all as they do not describe the role and responsibilities of the main contractor during the preconstruction period. The research by Barlow *et al.* found that it is 'possible that traditional models of the construction process are increasingly unable to fulfil new demands of clients, necessitating a search for alternative approaches'[164].

---

[161] NEDC (1991), 9.
[162] Thornton (2004), 3.
[163] Thornton (2004), 3.
[164] Barlow *et al.* (1997), 5.

## 3.9 New procurement procedures in place of gambling on incomplete information

There is a prevalent view that some clients will always believe they can identify consultants and contractors willing and able to take major financial risks in pricing a project. This gamble is intended to lock in a competitive price at the beginning of the construction process, but can often disregard the extent of the gamble taken by a consultant or contractor in pricing on this basis. Smith recognised the appeal of apparent cost certainty obtained through fixed price quotes, whatever the inherent risks, but went on to state that 'as a practical matter, things seldom work out this way'. He described this philosophy as 'legalised gambling' approach to contracting[165].

Problems in the construction industry by way of unpredictable outturn costs, delays and defects can often be traced to the single-stage, lump sum pricing process by which the contractor is expected to assess a correct market price for a project that it has not previously built on a site in respect of which there is little information, adopting a design which may still be incomplete or subject to revision and using a labour force and supply chain not yet recruited[166].

R.J. Smith recommended that clients should build more enlightened procurement and risk management procedures into their standard contract documents and suggested that those clients who adopt this approach have encountered fewer delays and disputes as well as establishing better relationships amongst project team members[167]. Significantly, he also suggested that such owners 'have obtained more competitive bids'[168].

This chapter has identified limitations in standard form building contracts that govern only the construction phase of a project. It has also established links between the single-stage procurement process that allows these limitations and the causes of typical claims and disputes. The next step is to contrast the single-stage procurement and contracting process with the features of a two-stage approach that facilitates earlier contractor involvement in line with Project Flowchart 2 set out at the end of this chapter.

---

[165] Smith R.J. (1995), 44.
[166] See Burke (2002), 237. The limitations and risks of single-stage tendering are considered further in Chapter 1, Section 1.2 (Early contractor appointments and project pricing) and Chapter 4, Section 4.3 (Preconstruction pricing processes).
[167] Smith R.J. (1995), 44.
[168] Smith R.J. (1995), 45.

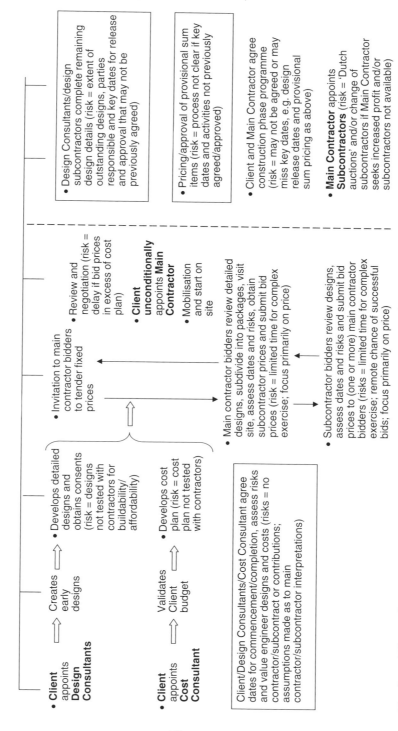

Project Flowchart 1 Typical single-stage construction phase contractor appointment

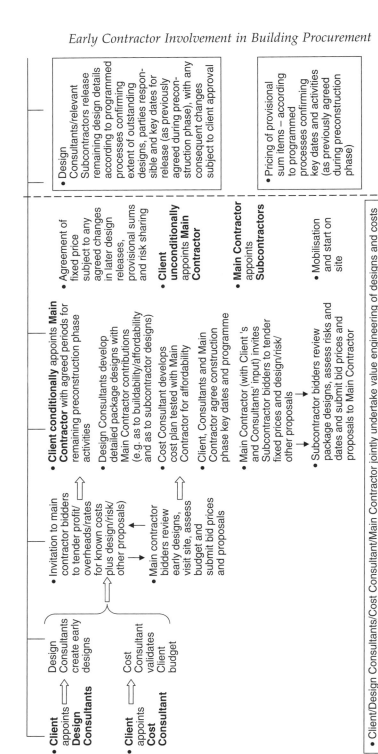

**Project Flowchart 2** Typical two-stage preconstruction phase and construction phase contractor appointments

# CHAPTER FOUR
# EARLY CONTRACTOR INVOLVEMENT IN DESIGN, PRICING AND RISK MANAGEMENT

## 4.1 Introduction

This chapter will explore the potential benefits to a project that can be achieved by means of early appointment of main contractors and certain subcontractors.

The argument for early appointment of the main contractor was articulated clearly by Banwell in 1964, who stated that there are 'occasions when it is appropriate for the main contractor to be appointed and brought into the team before the design is finished and the programme of work finally settled'[169]. In addition to early main contractor appointment, Banwell recommended the early appointment of certain specialist contractors as design team members.

The principle of the main contractor being 'brought into the team' is an interesting choice of words as many consultants and other commentators are still ambivalent as to whether the main contractor should be a fully-fledged team member or should remain at arm's length under procedures and contracts predominantly designed to transfer risk and maintain scrutiny of potential sharp practices. Appointment of a main contractor 'before the design is finished' recognises the contribution that main contractors can bring to the design of a project. Banwell goes on to state that 'Many general building and civil engineering contractors have developed highly specialised techniques in design and construction which can usefully be taken into account by the designer in formulating his scheme'[170].

The purpose of a preconstruction phase appointment is to obtain additional contributions to a project from a main contractor and from its subcontractors and suppliers. Project Flowchart 2, at the end of Chapter 3, sets out the procurement activities involved in a typical two-stage preconstruction and construction phase contractor appointment, noting the benefits that can be obtained and the risks that can be

---

[169] Banwell (1964), 4, Section 2.6.
[170] Banwell (1964), 4, Section 2.6.

reduced if contractors have an early involvement in project processes. These benefits include:

- Designs – designs can be developed with main contractor and specialist subcontractor contributions to establish their buildability and affordability at an early stage.
- Costs – the cost plan developed by the cost consultant can be tested for affordability with the main contractor and with subcontractor bidders at each stage of design development.
- Risks – the client's and consultants' risk assessments can be compared with those of the main contractor and its subcontractors, and risk management actions can be agreed and implemented without delaying start on site.
- Main contract tender – an early invitation to tender allows main contractor bidders to assimilate less advanced project information and propose improvements, with cost transparency achieved through pricing contractor profit, overheads and preconstruction phase costs. It also allows the client to obtain bids that demonstrate the contractor's design/risk/other contributions, and its ability to meet the client's budget, programme and performance standards.
- Subcontract tenders – the selected main contractor requires its subcontract bidders to assimilate complex project information at a stage when such information has already been agreed and has been set out in packages prepared and issued by the main contractor, allowing subcontractors to bid fixed prices and also to demonstrate their capability and their proposed design/risk/other contributions.
- Joint activities – time and processes can be created for joint client/consultant/contractor activities such as value engineering and joint risk management activities, and for the agreement of outputs from such activities, without delaying start on site.
- Programme – the construction phase programme can be agreed prior to start on site, including key dates for activities such as release of remaining consultant and contractor design details and the pricing and approval of provisional sum items.
- Subcontractor appointments – subcontractor appointments can be finalised by the main contractor prior to start on site (unless otherwise agreed by the client), creating greater cost certainty and greater subcontractor commitment.

The Arup report for OGC recognised the preconstruction phase appointment created under PPC2000 and observed that:

> 'The processes that incrementally develop the scope of the project provide for early involvement of the Contractor and encourage dialogue in open-book reporting of information. In doing so, the PPC

contract seeks to provide clarity and certainty of the details of the project and the responsibility of the parties in order to reduce the matters that could give rise to dispute'[171].

Leaving programming for further consideration in Chapter 5, the primary preconstruction phase processes that form part of two-stage procurement can be subdivided as three sequential but overlapping streams of activity, namely design, pricing and risk management. Each will be examined in turn, using empirical evidence from project case studies to illustrate the effect of these processes on project outcomes.

These project case studies are set out in Appendix A and are drawn from a number of sectors, project types and geographical regions:

(1) Bewick Court, Newcastle – housing refurbishment
(2) Watergate School, Lewisham – newbuild school
(3) Poole Hospital, Dorset – hospital refurbishment
(4) Bermondsey Academy, Southwark – newbuild school
(5) Macclesfield Station – rail refurbishment
(6) Nightingale Estate, Hackney – housing refurbishment
(7) A30 Bodmin/Indian Queens, Cornwall – newbuild road
(8) Project X – newbuild housing.

## 4.2 Preconstruction design processes

### 4.2.1 Contractor design contributions

The single style procurement approach to design development is to leave it in the hands of design consultants for as long as possible, on the basis that the greater the level of design detail when prices are invited from contractors, the greater the certainty of the prices quoted.

While recognising that the architect or engineer will generally act as design team leader, Banwell stated that 'it is essential in our view, and current experience bears us out in this respect, that the specialist consultants, some of whom may in fact also be specialist contractors, should be brought in at the earliest stage as full members of a design team'[172]. As to the appropriateness of the architect or engineer as leader of the design team, Banwell commented controversially that 'it is the quality of cooperation within the team rather than the identity of its leader that is really important'[173]. Banwell's argument was based on

---

[171] Arup 2008, 38.
[172] Banwell (1964), 4, Section 2.5.
[173] Banwell (1964), 4, Section 2.5. However, see also Appendix A, Project case study 8 regarding the lack of objectivity exhibited by an architect when acting as project manager in assessing the impact of late or incomplete design information.

the recognition that construction projects are increasingly complicated and highly mechanised, and that even the very best design consultants cannot have at their fingertips all the detailed knowledge available to specialist contractors through their research and development departments and through their on-site project teams.

The case for main contractor and subcontractor participation in design development is founded on the proposition that in many cases design consultants alone cannot develop designs that:

- Are sufficiently detailed to be capable of comprehensive fixed pricing by a main contractor in a single-stage tender;
- Are fully buildable by a main contractor without further detailing and/or amendment to reflect particular circumstances on site and the interaction between various trades;
- Embody the latest thinking of manufacturers, suppliers and specialist trades.

The JCT CE Guide states 'It is important that contractors and any key specialists are engaged early, ideally at a stage when the proposed design is not complete so that it is possible for the contractor and key specialists to consider ways in which the design can be made easier to build and maintain'[174]. It is therefore surprising that JCT CE is not itself a two-stage contract.

Banwell recognised in 1964 the need to appoint the main contractor 'before ... the programme of work [is] finally settled' and suggested that there is a role for the main contractor alongside the rest of the team in completing missing details before embarking on the construction phase of the project[175].

Banwell suggested that in many cases design consultants cannot be confident that their designs adopt the most practical and economical approach without the opportunity to review those designs in conjunction with the organisation or organisations that are going to implement them on site. For this to be done properly, and for cost and time benefits to be achieved, Banwell recognised the need for a significant period of time prior to commencement of the construction phase during which the main contractor is appointed to the team. While Banwell's recommendation of main contractor contributions to design might be interpreted simplistically as a case for 'design and build'[176], his proposal that the client should appoint the main contractor to assist in finally settling the programme of work is a clear argument for joint working during the preconstruction phase.

---

[174] JCT CE Guide, Section 38, 7.
[175] Banwell (1964), 4, Section 2.6.
[176] See Chapter 9, Section 9.2.2 (Design and build projects).

*Design, Pricing and Risk Management*

As Banwell observed:

> 'To call in a contractor to the site on which a complicated scheme – the planning of which may have taken many months or even years – is to be executed, and to expect him to be able to make himself thoroughly familiar with his task and to settle the right way in which to do it, when work must start within a few weeks or days, is unreasonable'[177].

Hence, Banwell's view was that in order to capture the skill and knowledge that an increasing number of contractors offer 'in, for example, deciding on types of construction to be adopted and the programme to be followed', it is important to decide correctly 'the point in time during the overall process of planning and construction at which the appointment of the contractor is made'[178].

> A30 Bodmin/Indian Queens (newbuild road): the early appointment of the main contractor allowed it to contribute to the buildability of the road, for example by influencing its route and the replacement of a viaduct with a steep-sided embankment[179].

The CIRIA report, *Selecting Contractors by Value*, recognised that contractors can make significant contributions by offering design alternatives that achieve the desired performance, or better, at less cost. The CIRIA report highlights the ability of contractors to identify opportunities for:

- 'Using more cheaply sourced materials;
- Installing components that match the form and/or function of those originally specified but at less cost, for example manufacturers' standard products in place of uniquely designed or specified items;
- Engineering the design for ease of construction without impact on aesthetics or fitness for purpose by making use of the contractor's particular skills or facilities;
- Precasting or pre-assembly;
- Increasing the degree of repetition between similar components;
- Reducing the need for temporary works'[180].

---

[177] Banwell (1964), 4, Section 2.6.
[178] Banwell (1964), 9–10, Section 3.13.
[179] Project case study 7, Appendix A.
[180] CIRIA (1998), 23, Section 2.4.

*Early Contractor Involvement in Building Procurement*

These contributions can be made either by main contractors or by subcontractors or suppliers, depending on the features of the relevant project. The National Audit Office, in its case study of office relocation by the Department for International Development, referred to early involvement of the main contractor under a two-stage approach allowing 'the design brief to be developed to take into account the scope of the changes required'[181]. It also recognised that this two-stage approach enabled the client 'to involve subcontractors at an early stage to identify the most effective way of progressing the project'[182].

> Poole Hospital, Dorset (hospital refurbishment): the client recognised the importance of ensuring that mechanical and electrical equipment in its operating theatres achieved its performance specifications. It appointed its main contractor and specialist subcontractors early so that they could contribute to finalising design details for such equipment and to the logistics governing its integration with existing facilities in a restricted clinical environment[183].

### 4.2.2 Integration with consultant designs

Where preconstruction design development is led and controlled by architectural and engineering consultants, contractor involvement in such design is often postponed until a late stage under the single-stage procurement approach and this remains the default position in most standard forms of appointment published by consultant professional bodies. For example, RIBA (2004) and ACE (2002) did not reflect Latham's concern that construction contracts, where 'design and construction are totally separated, in that the main contractor and subcontractors have no design responsibilities or involvement in the preparation of the design', do not 'relate easily to the reality on modern construction sites'[184]. Design as a preconstruction phase process was set out in RIBA (2004) through stages 'A/Appraisal, B/Strategic Briefing, C/Outline Proposals, D/Detailed Proposals, E/Final Proposals, F/Production Information, and G/Tender Documentation'[185]. These first seven stages made no provision for main contractor or subcontractor contributions and the only reference to contractor design input was at Stage 'K/Construction to Practical Completion', where the architect is required to 'Review design information from contrac-

---

[181] NAO (2005), Case studies, 24.
[182] NAO (2005), Case studies, 25.
[183] Project case study 3, Appendix A.
[184] Latham (1994), Section 5.17, 36.
[185] RIBA (2004), Schedule 2, Services, 4.

*Design, Pricing and Risk Management*

tors or specialists' during the construction phase[186]. The detailed sequence of activities set out in the RIBA Plan of Work published in 2008 created greater flexibility but still assumed no contractor involvement in design prior to the construction phase of the project other than by the contractor (acting as client) in obtaining the Contractor's Design Services from consultant architects and engineers in relation to a 'Design and Build' procurement model or a 'Contractor's Designed Portion'[187]. By contrast, the ACE (2002) conditions provided for the possibility that 'the detailed design of any part of the Works should be carried out by a Contractor or Sub-Contractor'[188].

N.J. Smith observed that 'There is ... a likelihood of designers in these disciplines acting in a "tribal" manner'[189] and it is of course arguable that tribal behaviour influences the apparent divide between design consultants and contractors.' N.J. Smith suggested that 'experience shows that this tribal behaviour, if effectively managed, can lead to significant innovations in the design stages of projects, and does not necessarily have a negative impact on the organisation of the design'[190]. The need to manage tribal behaviour between design consultants and contractors would justify, as a starting point, the clearer alignment of the preconstruction phase design processes described in consultant forms of appointment[191]. More fundamentally, it also underlines the need to bridge the lack of integration between design consultants and the design activities of main contractors and subcontractors.

Main contractors have contributions to offer in terms of assessing the affordability and buildability of designs. Specialist subcontractors and suppliers have extensive knowledge of particular design functions, for example in the field of mechanical and electrical engineering. If a standard form building contract does not recognise the potential for such design input, then the client is not taking full advantage of the contractor's capabilities as a designer. Further, by retaining control through its consultants of the design process, the client is also retaining the risk of the affordability and buildability of the designs[192]. Thus, in excluding or postponing main contractor and subcontractor design contributions, the client is not only missing the potential for added value but is also

---

[186] RIBA (2004), Services Supplement: Design and Management A.
[187] RIBA 2008, 69–91.
[188] ACE (2002), B2 (Obligations of the Consultant), clause 2.7.
[189] Smith N.J. (2002), 243.
[190] Smith N.J. (2002), 243, 244.
[191] The Construction Industry Council published in 2007 an integrated set of consultant appointments, CIC (2007). The 2008 RIBA Plan of Works describes in parallel the services of the lead designer, architect, structural engineer and services engineer at each stage of the project, RIBA (2008), 20–47.
[192] See Cox & Thompson (1998), 93, as to client liability for consultant designs without contractor input.

increasing its own liability to the main contractor in respect of consultant designs.

> Watergate School, Lewisham (newbuild school): the client expected to novate its own design consultants to the appointed main contractor, but nevertheless invited bidders to propose their own alternative design solutions. The successful bidder put forward its own proposed design solution, and appointed its own design sub-consultants in place of the client's design consultants, and accepted full responsibility for achieving that solution[193].

### 4.2.3 Integration with subcontractor and supplier contributions

In many respects, the main contractor is primarily a manager of the subcontractor and suppliers that contribute to a construction project. Significant design contributions are more likely to be made by specialist subcontractors (for example, mechanical and electrical subcontractors) or by specialist suppliers (for example, cladding or plant manufacturers). One option, therefore, is for the client and its consultants to consider early appointments of key specialist subcontractors/suppliers so as to allow them to participate in design development.

The importance of early subcontractor and supplier appointments is emphasised by the Institution of Civil Engineers, quoted in *Accelerating Change*: 'Designers must involve the contractors, specialist subcontractors and key manufacturers as soon as possible. In order to interpret and develop a functional brief it is essential that designers (including specialist subcontractors and key manufacturers) are able to get close to clients'[194].

An attraction of early subcontractor and supplier appointments is that relevant design consultants can deal directly with the corresponding specialist subcontractors and suppliers can access their research and can extract added value proposals. The early appointment of subcontractors and suppliers was clearly endorsed in the 2005 National Audit Office Report which expresses continuing concern at the lack of evidence that public sector clients are 'involving specialist suppliers, such as mechanical and electrical engineers, as fully and early as they might despite the fact that these suppliers are often critical to the delivery of successful construction projects'[195].

---

[193] Project case study 2, Appendix A.
[194] Strategic Forum (2002), Section 26.
[195] NAO (2005), 54.

*Design, Pricing and Risk Management*

> Nightingale Estate, Hackney (housing refurbishment): the architect and structural engineer were leading practices with strong views as to design solutions. However, they recognised the complexity of the technical issues surrounding concrete repairs, the relocation of mechanical and electrical risers, and the need to divide a mid-rise residential block into two buildings. The early appointment of the main contractor under a preconstruction phase agreement created an opportunity to test the buildability and affordability of consultant designs and to obtain new ideas from specialist subcontractors. The advice of the concrete specialist was required to ascertain what repairs were needed and what timescale should be allowed in the construction phase programme. Leonard Stace, cost consultant on the Nightingale Estate project, observed: 'We needed concrete specialists on board early in case we had to rejig the programme: they would know whether these changes were achievable. A close-knit team with a common objective … was the only way this approach could work'[196].

For the client and its consultants to develop designs with subcontractors and suppliers in the absence of a main contractor there are a number of potential disadvantages:

- The main contractor, when appointed, may object to the pre-selected subcontractor or supplier and put forward a convincing case as to why its own alternative subcontractor or supplier would be more appropriate.
- Design proposals conceived by design consultants and subcontractors or suppliers still need to be integrated with the design and construction of the project as a whole. The main contractor's appreciation of overall buildability will still remain to be taken into account once it is appointed, and this could challenge design assumptions that by then have been well advanced by consultants, subcontractors and suppliers.
- The relationships between the main contractor and its proposed subcontractors and suppliers need to be finalised. Subcontractor or supplier design proposals will need to be integrated with the main contractor's programme and with the overall responsibilities attributed to the main contractor under the client's brief.

---

[196] Project case study 6, Appendix A.

- Where the client and its consultants have progressed subcontractor or supplier appointments in advance of main contractor involvement, the main contractor may assume incorrectly that the client has dealt with related subcontractor and supplier interfaces (e.g. relationships between surveyors, suppliers and installers of specialist equipment). As a result, the main contractor may neglect or delay dealing with interface issues that would otherwise have been picked up under its own supply chain procurement procedures.

Accordingly, even where design input can be identified as primarily available from a limited range of specialist subcontractors and suppliers, there are significant advantages in the early appointment of the main contractor as well as the early appointment of those subcontractors and suppliers.

> Project X (newbuild housing): the design for the project was conceived by the specialist manufacturer of residential modular units, and was further developed by the consultant architect and engineer. The main contractor, when appointed at a later stage, took a relatively passive role in design development and integration, and then used consultant design errors, variations and delays as the basis for time and money claims[197].

### 4.2.4 Payment for contractor design contributions

If a contractor is appointed to work on preconstruction designs alongside the client's consultants, it is likely to expect payment. A 1975 NEDO report specifically envisages that 'More formal inducements can be provided in the form of consulting fees to the contractor or schemes for sharing cost savings between the client and contractor'[198]. On a cautionary note, the 1975 NEDO report also recognises that 'Collaborative working with the contractor can impose additional work on the design team. This should be allowed for in the programming of work for an in-house design team or in the fees of consultant designers'[199]. Additional costs should not arise if preconstruction phase design contributions are clearly described in an agreed sequence. However, the later contributions of new parties and the reworking of designs to reflect new information need to be carefully managed so

---

[197] Project case study 8, Appendix A.
[198] NEDO (1975), Section 7.24, 73.
[199] NEDO (1975), Section 7.25, 73.

that the parties' commitment to methodical design development and creative value engineering does not obligate them to duplicate, without further payment, design activities that they have already undertaken for an agreed fee.

Cost is likely to affect the enthusiasm of clients for early main contractor appointments, particularly if they are asked to pay not only main contractor design fees but also additional consultant design fees to accommodate joint working with the main contractor. It is important, therefore, to establish clearly the benefits that the client will receive as a result of paying such fees, namely the contributions made by the main contractor in terms of deliverables described in the preconstruction phase agreement, with related intellectual property rights as appropriate[200].

> Watergate School, Lewisham (newbuild school): the main contractor and the architect/engineer agreed fixed fees for all work during the preconstruction period, including the value engineering necessary to achieve prices within the client's budget. These fees were not exceeded[201].

> Nightingale Estate, Hackney (housing refurbishment): the preconstruction period was extended to allow for the additional consultant design work required to implement alternative solutions proposed by the main contractor and its specialist subcontractors. Additional consultant fees were negotiated and agreed by the client in advance, and therefore could be taken into account in minimising any impact on the overall project budget[202].

As to the cost benefits of obtaining contractor design contributions, the 1991 NEDC report found that working together with the contractor earlier than usual during the design phase 'has reduced the costs associated with rework due to poor understanding of the client's requirements'[203].

---

[200] PPC2000 provides for the grant to the client of a licence in respect of intellectual property rights in all designs produced by the contractor, PPC2000 Partnering Terms, clause 9.2.
[201] Project case study 2 Appendix A.
[202] Project case study 6, Appendix A.
[203] NEDC (1991), 78.

## 4.3 Preconstruction pricing processes

### 4.3.1 Contractor contributions to finalising prices

A client's choice of its procurement and contracting strategy will be significantly influenced by pricing. Nearly all projects have budgetary constraints and it is understandable that the client will wish to convert its budget into a fixed price at the earliest opportunity. However, this can lead a client and its advisers to require the main contractor to commit itself to a fixed price before it has sufficient information as to designs, supply chain members and risks to be able to make an accurate calculation[204]. The JCT CE Guide states that:

> 'Under traditional single stage tendering arrangements the opportunity to plan for the construction stage is restricted ... [contractors] will do enough preparatory work to be successful at tender but are unlikely to be able to understand fully all aspects of the project or have sufficient time to identify and consider how to manage the potential risks to the project'[205].

This in turn can lead bidding main contractors to add arbitrary contingencies or premiums to their quoted prices, with the following possible consequences:

- A windfall by way of additional profit for the main contractor, and therefore potentially wasted money for the client, if the contingency is higher than necessary to cover the main contractor's costs; or
- A loss for the main contractor, resulting in pressure to make additional claims on the client, if the contingency is insufficient to cover the main contractor's costs.

Either the client or the main contractor can lose out as a result of guesswork by the main contractor in pricing a project. To avoid this, either the client needs to obtain detailed designs and risk assessments from its consultants sufficient for a main contractor fully to assess its position and quote a fixed price without significant contingencies or premiums, or the client needs to involve the main contractor in a joint process governing the build-up of prices.

---

[204] See also Chapter 3, Section 3.3 (Standard forms and the assumption of complete information) regarding the treatment of the construction process as the purchase of a product. See also Chapter 9, Section 9.2.4 (Single-stage tendering) regarding the weaknesses of single-stage tendering.
[205] JCT CE Guide, Section 35.6.

## Design, Pricing and Risk Management

### 4.3.2 Information required for accurate pricing

The agreement of an accurate price is dependent on all relevant information being held both by the client requesting the price and by the main contractor providing the price. Otherwise, information held only by one party or the other can distort the price quoted, for example information withheld by the client as to the obstacles that the main contractor will encounter on site, or information withheld by the main contractor as a risk contingency quoted in respect of possible obstacles on site[206]. If one party withholds information as a result of which another party is placed under an excessive obligation or is deprived of a reasonable choice, then this does not create the foundations for a successful working relationship. If, however, the maximum information is made available to all parties, for example, as to the underlying costs of a project, then this can contribute significantly to the efficiency of later working relationships – for example in managing change and minimising disputes in relation to the cost effects of matters outside the parties' control[207].

> Macclesfield Station (rail refurbishment): the client had faced cost overruns on previous projects due to main contractor claims based on incomplete consultant designs and undisclosed risks. The client appointed the main contractor early on the basis of agreed profit and overheads, so as to create under a preconstruction phase agreement a preparatory period during which the team could develop full information necessary for them to commit to a fully designed and fully costed project in advance of start on site[208].

### 4.3.3 Adjustment of prices to reflect new information

The parties need ways of adjusting their original financial arrangements to reflect new information acquired during the course of their relationship. Rather than fixing prices based on incomplete information, it may be preferable for the parties to agree 'incentive-efficient mechanisms'[209] to determine future prices in a way that will achieve

---

[206] See Chapter 3, Section 3.8 (Links between claims and building contracts) regarding the risks perceived by Judge Anthony Thornton deriving from failure to exchange information that leads to a lack of trust 'at every level of the design, planning and construction chain', Thornton (2004), 3.
[207] See Milgrom & Roberts (1992), 140, as to 'informational asymmetric'.
[208] Project case study 5, Appendix A.
[209] Milgrom & Roberts (1992), 143.

more accurate figures and will offer better value and will still protect their reasonable interests and expectations.

A well-established approach is the use of 'provisional sums'[210], whereby an approximate figure appears in the contract price in respect of an element of the project for which there is insufficient information to agree a fixed price. The provisional sum is then converted to a fixed amount during the construction phase by means of a procedure of design development and pricing of the developed design (possibly by subcontract tendering), leading to an instruction to the main contractor once there is sufficient price certainty for the relevant item to be authorised. This approach postpones the client's cost certainty in respect of a provisional sum item until completion of a process that takes place after the client has made an unconditional commitment to the project. There is, therefore, a contractual commitment of the client and main contractor to joint pricing of the relevant item.

Under a two-stage procurement model, equivalent joint pricing processes can be implemented with the main contractor to price works packages as sufficient design detail is developed during the preconstruction phase[211]. Subject to the agreed sequencing of consultant design releases, contractor design contributions and joint subcontractor tendering, it is possible by this means to achieve fixed prices for such works packages ahead of start on site[212].

### 4.3.4 Prices and contractor selection

The extent to which pricing is used in the selection criteria for main contractors will depend on the extent to which the client can be confident that bidders have enough information to tender the best value prices for the project, and the extent to which the client wishes to obtain added value from the selected main contractor in the development and finalisation of its prices. There are certain types of project where designs are straightforward, with limited scope for main contractor input, and where the components of the main contractor's supply chain can be easily identified and priced, for example if they involve supply of standard components that are likely to be subject to pre-existing long-term arrangements between main contractors and their subcontractors or suppliers[213]. In these circumstances, the selection of main contractors

---

[210] For example, JCT 2005 SBC/Q clause 3.16.
[211] See Chapter 4, Sections 4.3.5 (Two-stage pricing), 4.3.6 (Treatment of profit and overheads) and 4.3.7 (Concerns as to two-stage pricing) regarding the details of joint pricing processes.
[212] See Chapter 5, Section 5.4 (The role of binding programmes) as to the basis for agreeing a clear sequence of binding deadlines.
[213] See Chapter 9, Section 9.2.1 (Size and simplicity).

at least in part according to fixed price quotes is a practical proposition, as bidders can assess the implications of the designs provided to them and can establish robust prices among the members of their supply chain.

However, for projects that involve more complex designs to which not only the main contractor but also key subcontractors and suppliers could contribute, and for projects where pre-existing main contractor supply chain arrangements cannot be assumed to exist as a source of robust tender prices, the requirement for comprehensive fixed price quotes as a basis for main contractor selection may have the following adverse consequences:

- Inflation of the prices quoted to cover perceived main contractor risks, with the difficulty for the client of then identifying and challenging the pricing of a bidder's risk assessments after selection if the client wishes to negotiate reduced risk contingencies.
- Limited main contractor enquiries among its supply chain during the tender period, with the consequence of estimated supply chain price components adding up to an estimated main contract tender price. This could be unnecessarily high as a result of risk contingencies added by the main contractor or any of its subcontractors and suppliers, or could be unrealistically low as a result of the main contractor or any of its subcontractors and suppliers submitting over-optimistic estimates in order to win the job.
- Reluctance of the selected main contractor and its subcontractors and suppliers to participate in the subsequent development or refinement of designs, for fear of the client expecting design enhancements without being willing to pay for them. For example, a NEDO case study concluded that 'The contractor was involved in the design stage, but he found it led to numerous small disputes on detail where the client wanted more expensive solutions to the specification. Such increased costs fall entirely on the contractor's profit margin if the price is already fixed'[214].

Banwell observed that in the private sector he was aware of projects 'in which the contractor has been appointed early to work as part of the team in developing the details of the project and establishing its cost'[215]. In supporting this approach, Banwell challenged the widespread assumption that the client's interests are best served by single-stage tendering leading to a contract under which the main contractor is appointed only after a lump sum price for the project has been

---

[214] NEDO (1975), Case study, Table B1, 116.
[215] Banwell (1964), 10, Section 3.13.

agreed. Banwell noted that in almost all cases of which he was aware public sector clients had not appointed the main contractor early, and he attributed this to fears that 'to do so would be contrary to established notions of public accountability'[216].

### 4.3.5 Two-stage pricing

Two-stage pricing commences with early main contractor selection as the first of two stages of competitive tendering, establishing the main contractor's profit, overheads, preconstruction phase costs, approach to risk pricing and any other cost components that can be priced accurately by bidders on the basis of the design and other information available at that first stage.

Main contractor selection should be based on a wider range of criteria than such pricing elements alone, for example those outlined in the CIRIA report *Selecting Contractors by Value*:

- 'Technical knowledge and skills – experience in engineering specialist elements; appropriate design capacity;
- A number of management skills: ... managing time ... managing cost ... managing value ... managing quality ... managing risk ... managing health and safety ... ;
- Effective internal organisation – clear communications; sound administration; empowered staff;
- Collaborative culture – record of 'partnering'; positive lead from the top; client focus;
- Appropriate human resources – qualified and enthusiastic personnel available to do the job;
- Supply chain management – sound dealing with subcontractors/suppliers; established relationships;
- Financial resources – sound balances and cash flow; reliable references;
- Generally – a sound, relevant and demonstrable track record'[217].

Such criteria are more demanding for the client and its consultants to assess than a straightforward comparison of prices, but they should yield valuable information that will assist the client in making the right choice and that will be consistent with two-stage procurement.

---

[216] Banwell (1964), 10, Section 3.13. See also Chapter 9, Section 9.3.1 (Constitutional or regulatory constraints).
[217] CIRIA (1998), 15.

> Watergate School, Lewisham (newbuild school): the client assessed main contractor bids on the basis of criteria that comprised innovative design solutions, proposals for working with stakeholders and managing supply chains, proposals for site welfare and quality control, and quoted lump sums for design fees, profit and overhead[218].

The main contractor's appointment is then followed by the second-stage competitive selection of subcontractors and suppliers, undertaken by the main contractor working with the client and consultants. This second-stage pricing process can take advantage of more complete information established by means of the main contractor's input to design, risk management and programming.

Two-stage pricing was already contemplated in the Banwell report as a means to combine early main contractor selection with competitive processes. The two-stage procedure envisaged by Banwell began with selection of the main contractor through 'a preliminary competition based on an outline, in which the offers of selected firms are considered in the light of such factors as management and plant capacity, and the basis of their labour rates, prices and overheads'[219]. Banwell then envisaged a second-stage procedure during which 'the chosen contractor works as a member of the team, while details are developed and bills of quantities drawn up, and at the end of this time submits a more detailed price which if satisfactory becomes the formal contract sum'[220].

Two-stage tender procedures were also recognised in the National Joint Consultative Committee (NJCC) Code of Procedure for two-stage selected tendering. However, the NJCC procedure did not envisage any contractual relationship between the client and the main contractor until after the second-stage process of finalising subcontract packages and other costs had been completed. This fails to acknowledge the investment required by the main contractor in the second stage of the two-stage process, and the advantages to both the client and the main contractor if this process is governed by a preconstruction phase agreement.

Banwell specifically contemplated a conditional preconstruction phase agreement as the basis for the main contractor's first-stage appointment, not least so that the client can stand down the original contractor and select an alternative in the event that a satisfactory price does not emerge. He perceived that this two-stage arrangement would

---

[218] Project case study 2, Appendix A.
[219] Banwell (1964), 10, Section 3.14.
[220] Banwell (1964), 10, Section 3.14.

provide 'competition in a new sense' which will also 'enable the contractor to join the team at a time which is precluded by existing procedures'[221]. In Banwell's view, two-stage tendering in this clearly structured form would lead to 'undeniable advantages for the client in solving some of the failures in communication and understanding between designers and contractors and contractors and subcontractors which have hampered the industry in recent years'[222].

A two-stage pricing process has the further significant advantage of reducing the overall number of tendering exercises conducted in the marketplace. Instead of prospective subcontractors and suppliers each wasting time and money bidding to one or more of a number of main contractors who are themselves bidders (with that wasted cost recovered in the subcontractors' and suppliers' prices on the projects where they are successful), they will be in the position of bidding to a pre-selected main contractor. This significantly increases their chances of success and their likely commitment to the tender process.

Two-stage pricing is possible whatever the pricing model adopted for payment of the construction phase contract price. This could be lump sum or remeasureable, with payments calculated according to a bill of quantities or schedule of rates. Equally, it could be cost reimbursable in its entirety or within the limits of a target cost with agreed shares of any excess or shortfall. Any of these options can be used as appropriate in the second-stage tendering of subcontract packages. However, remeasurable second-stage pricing is likely to be more appropriate for civil engineering works such as roads[223]. Cost plus second-stage pricing will only be appropriate where the client has other means of motivating the main contractor to minimise its expenditure[224].

### 4.3.6 Treatment of profit and overheads

A significant benefit of the two-stage pricing approach is that the client can obtain a clear understanding of the main contractor's expected return from the project, by way of both profit and overheads. By identifying and agreeing these amounts in advance, they can be distinguished from the underlying costs that both the client and the main contractor may need to reduce so as to achieve a price within the client's budget. Agreement of profit and overheads also obtains for the client greater control over otherwise invisible benefits that the main

---

[221] Banwell (1964), 10, Section 3.15.
[222] Banwell (1964), 10, Section 3.15.
[223] See for example Project case study 7, Appendix A.
[224] See for example Project case study 10, Appendix A.

*Design, Pricing and Risk Management*

contractor might traditionally expect through practices such as supplier rebates and prompt payment discounts on amounts that are contractually due to subcontractors and suppliers[225].

Agreement of profit and overheads separate from underlying costs on an open basis is a significant contributor to managing the cost of change and to understanding and agreeing the cost implications of events of delay and disruption. It can therefore help the parties to avoid or minimise the scope for disputes. Arup observed in relation to PPC2000 that:

> 'When differences arise against a background of open-book record-keeping and the cooperative exchange of information, the process and disclosure of information can reduce the scope of the difference. This is supplemented by the requirement to give early warnings. The prospects of a difference becoming a dispute can be mitigated by enabling a focus on the specific issues between the parties and how these tie it to their objectives'[226].

Two-stage pricing can further align the client's interests with those of the main contractor if profit is fixed as a lump sum rather than a percentage, so that reduced costs will not reduce the main contractor's pre-agreed return from the project. This approach can be extended to incentivisation arrangements by increasing the profit to the main contractor (and possibly other team members) where agreed savings are achieved.

> Watergate School, Lewisham (newbuild school): the main contractor's profit and overhead, and its design sub-consultants' fees, were agreed as fixed amounts irrespective of any increase in the cost of the project, thus ensuring that they would not gain from any increase in other project costs[227].

### 4.3.7 Concerns as to two-stage pricing

The process of two-stage pricing not only requires a more demanding contractor selection process to be implemented by the client but, as CIRIA observed, 'can also place considerable demands on contractors', and two-stage pricing therefore needs to be 'carefully thought out,

---

[225] For example, PPC2000 Partnering Terms, clause 12.8 prohibits discounts or other benefits payable by any subcontractors or suppliers to the main contractor unless approved in advance by the client.
[226] Arup (2008), 50.
[227] Project case study 2, Appendix A.

to balance the expenditure of resources against the benefit to be obtained'[228]. For example, a contractor needs to prepare the qualitative proposals required to meet the criteria for its own selection in the absence of a fully priced bid and, if successful, needs to organise its own supply chain selection procedures and criteria so as to implement second-stage procurement processes.

From the main contractor's point of view, there will be additional concerns that a two-stage pricing approach may involve it in a protracted period of working with the client without remuneration and without a clear understanding as to the terms on which its appointment will be confirmed and when this will occur[229]. The chief QS of a leading construction company observed that:

> 'Two-stage tendering is now commonplace in design and build, but we still see projects with no structured approach to the preconstruction activities ... Experience has taught us either:
>
> - The notional preconstruction period quoted is often significantly exceeded, leading to unrecoverable staff costs; or
> - A bespoke agreement is introduced later, when we are already committed and it is difficult to withdraw'[230].

The only means to avoid such uncertainties is to agree a series of binding deadlines that apply to the client and consultants as well as to the contractor. This supports the case for a preconstruction phase agreement which establishes a clear understanding as to the preconstruction phase tasks that the contractor will be performing in finalising the remaining costs and the duration of the preconstruction period during which this will occur. To deter clients from wasting a contractor's time in speculative preconstruction phase work, Banwell recommended a provision 'in the conditions of the preliminary appointment to pay the original contractor for work done during the working up period should his final price prove unacceptable'[231].

Concerns may be expressed that a pre-selected main contractor will be less commercially rigorous in its subcontractor and supplier tender procedures, and that inflated subcontractor and supplier prices could lead to the total price exceeding the client's budget. These concerns can be addressed by:

- Close monitoring of the main contractor's subcontract tendering procedures and documentation by the client and its consultants, to

---

[228] CIRIA (1998), 8.
[229] See also Chapter 9, Section 9.3.4 (Concerns as to conditionality).
[230] Project case study 8, Appendix A.
[231] Banwell (1964), 10, Section 3.14.

ensure that these do not impose excessive demands that could inflate subcontract prices;
- Establishment of a clear preconstruction phase agreement under which the main contractor and the remainder of the team are obliged to undertake value engineering of subcontractor and supplier prices so as to achieve a total price within the budget as a precondition for the construction phase of the project to proceed.

> Watergate School, Lewisham (newbuild school): the prices obtained by the main contractor from certain subcontractors and suppliers exceeded the amounts allowed in the client's cost plan. The main contractor and its design sub-consultants honoured their contractual commitment that they would, without additional payment, seek savings through value engineering sufficient to achieve a construction phase fixed price within the client's budget[232].

The two-stage pricing system is vulnerable when the main contractor is pricing any elements of the project that it wishes to undertake utilising its own labour, and any elements where it wishes to put forward a single preferred subcontractor or supplier with whom it has an established relationship. In either case, there could be cost benefits to the client but there is no obvious way of verifying such benefits without a second-stage competitive process unless the client is able to rely on advice from its cost consultant as to whether a direct labour proposal or a single source proposal offers value for money.

PPC2000 envisages that in these circumstances the main contractor will put forward a 'Business Case' as a means of establishing the benefits of a 'Direct Labour Package' (defined as any part of the project undertaken by the main contractor using its own direct labour) or 'Preferred Specialist' (defined as any subcontractor or supplier proposed by the main contractor in a business case to justify client approval without market testing). If the client on advice is not satisfied with such business case, then the main contractor will be obliged to implement a second-stage subcontract competitive tender process in respect of the relevant works package as a means of testing its business case against alternative subcontractor or supplier proposals[233].

Another area where two-stage pricing is vulnerable is in respect of risk contingencies. Although preconstruction phase joint risk management processes[234] allow the opportunity for the client, main contractor and other team members to analyse risks and seek ways of reducing

---

[232] Project case study 2, Appendix A.
[233] PPC2000 Partnering Terms, clause 10.
[234] As described in Chapter 4, Section 4.4 (Preconstruction risk management processes).

or eliminating them, it remains possible that a risk which the client wishes the main contractor to bear will continue to attract a cost within the contract price that the client considers excessive. The means of tackling this problem through a preconstruction phase agreement include:

- A clear understanding as to the main contractor's approach to risk pricing at the point of its selection, so that additional risk contingencies are not introduced at a later date during the preconstruction phase;
- A clear risk management system under which items costed in a risk register are subject to agreed activities intended to reduce such costs, undertaken in accordance with an agreed timetable;
- An incentive system for the main contractor (and possibly other team members) whereby profit and overhead are not reduced if a risk contingency is reduced or eliminated, and/or whereby an additional amount may be payable if a risk contingency is reduced or eliminated and/or no greater profit and overhead (or possibly a reduced amount by way of pain share) will be payable if a risk contingency is exceeded.

## 4.4 Preconstruction risk management processes

### 4.4.1 Early risk management

Risk has been defined as 'the possible adverse consequences of uncertainty'[235]. The identification of potential areas of risk and uncertainty, and careful consideration as to the appropriate ways to manage them, are an important feature in the creation and development of a project procurement strategy. This in turn begs the question of who should establish the best ways to manage risk and uncertainty, at what stage of the project and under what contractual systems.

Ways of reducing risk include:

- 'Obtaining additional information;
- Performing additional tests/simulations;
- Allocating additional resources;
- Improving communication and managing organisational interfaces'[236].

What Smith *et al.* do not indicate is when these activities should be undertaken in order to achieve the desired risk reduction. Clearly, they

---

[235] See Smith N.J. (2002), 193.
[236] Smith *et al.* (2006), 88.

*Design, Pricing and Risk Management*

will have greatest impact if undertaken during the preconstruction phase of the project.

Burke states that a range of responses to risk (to eliminate it, mitigate it, deflect it or accept it) 'should be developed in advance during the planning stage' of a project[237]. Although Burke clarifies the timing of risk management, he does not explore by whom his range of responses should be developed. It is suggested that at the planning stage of the project the main contractor should be appointed to work with the client and consultants in development of appropriate responses to risk.

> Macclesfield Station (rail refurbishment): joint risk management was essential to ensure that the main contractor's price and the agreed construction phase operations at the railway station took account of the required rail regulator approvals and the very limited periods available for work on site during station and line closures. Kevin MacConville, who worked on Project case study 5 as project manager stated that risk management: 'is a significant area where open and frank discussions and workshops can really drive and control the project objectives and ensure value for investment is achieved … The client has a most important role to play in this process – he needs to identify and explain what risks he is willing to undertake and for how long'[238].

### 4.4.2 Separate or joint risk management?

In a single-stage procurement model, risk assessment and analysis, and any consequent early risk management activities, are preparatory processes undertaken by consultants on behalf of the client without the involvement of the main contractor or any subcontractors or suppliers, on the assumption that bidding contractors will undertake their own separate risk analyses and risk management activities for their own reasons. In such cases, the consultants' risk assessment is interpreted by each bidding main contractor when formulating its bid price for the project, which may in turn give rise to negotiations between the client and the successful main contractor as to whether such interpretation is correct and whether the risks allowed for in such price are appropriate.

A view of single-stage risk management was expressed by Smith *et al.* as follows:

---

[237] Burke (2002), 239.
[238] Project case study 5, Appendix A.

'Risk management is undertaken by both client and contractor organisations, but for different reasons. Clients will usually be concerned with the best use of their capital resources, the likely cost of procuring the facility and their return from their capital investment. Contractors will be concerned with the decision as to whether to tender for a given project in terms of the returns obtainable, the desired competitiveness of their tender, and the most profitable means of constructing or increasingly designing and building the project'[239].

However, the activities that Smith *et al.* attributed to the tendering main contractors are simply a commercial risk assessment based on the information provided with a view to achieving the most profitable results, not a risk management process at all. Smith *et al.* focused primarily on risk assessment (although they called it risk management) in the context of the main contractor submitting a competitive bid. They did not acknowledge the comprehensive risk management exercises that have to be undertaken by the main contractor who goes on to win that bid and to take a transfer from the client of the typical range of risks involved in constructing and completing the project.

A single-stage process does not allow any bidding contractor (or its subcontractors or suppliers) to participate in the client or consultant risk assessment, nor does it allow a period of time for the client, its consultants and the main contractor (and potentially its subcontractors and suppliers) to work together to manage risks and reduce the prices attached to those risks. In addition, where a prospective main contractor is still in competition while priced risks are being negotiated, any attempt at joint analysis of such risks will be affected by the balance of negotiating power between the parties at that time[240]. It is also likely that each side will assume that the other's apparent perceptions of risk are in fact techniques to gain a more favourable financial position[241].

In a project where the main contractor has no right or responsibility to analyse risk jointly with the client, these considerations could lead the main contractor to a cynical assessment of the client's risk assumptions, the exploitation of any weaknesses in the client's documents and the use of the construction contract to secure additional profit by means of claims arising from risks which the client has not comprehensively transferred. This is a familiar scenario in disputes on projects procured

---

[239] Smith *et al.* (2006), 93, 94.
[240] See also Chapter 2 Section 2.6 (Limited efficiency caused by unknown items) regarding the impact of coercion on efficiency.
[241] See also Chapter 2 Section 2.7 (Risk and fear of opportunism) regarding the adverse effects of fear of opportunism.

*Design, Pricing and Risk Management*

through single-stage tenders and is in contrast to the cooperative approach that can be achieved by joint risk management.

Joint risk management is an important tool available to the project team to obtain recognition of other parties' actual or perceived risks and to agree the best way to deal with them.

> Nightingale Estate, Hackney (housing refurbishment): the main contractor acknowledged that its work on joint risk management, pursuant to a preconstruction phase agreement, identified serious issues affecting the structure of the block and requiring redesign. If the main contractor's input had not been invited and if the risk issues had not been addressed until the construction phase, the main contractor indicated that the cost and time required for redesign would have given rise to claims in excess of £500,000[242].

The client and consultants have considerable time to organise and assess their risks, whereas a bidding main contractor has only a period of a few weeks to undertake its separate assessment while at the same time compiling all other aspects of its response to the client's invitation to tender. The tendering main contractor's risk assessment is, therefore, a much abbreviated exercise bound up in its response to the technical and pricing requirements of the bid. Main contractors also need to pass many risks down to their subcontractors and suppliers, and might be more motivated to offer helpful risk management proposals if they were not required at bid stage to assess and absorb risks (as predetermined by the client and its consultants) within a fixed price bid not fully tested with their subcontractors and suppliers.

These constraints demonstrate the benefits of early contractor appointment as a means to enable a period and process of joint risk management. It is interesting to note the view of Bennett & Pearce that risk management will be more successful if the whole project team is appointed at an earlier stage in the project and that these early appointments should include early appointment of the main contractor. Bennett & Pearce stated that: 'The earlier the whole project team is appointed the better the risk management process will be. Contractors, consultants and other key suppliers bring knowledge and experience of construction, delivery and related financial risks that are helpful in managing risks'[243].

---

[242] Project Case Study 6, Appendix A.
[243] Bennett & Pearce (2006), 249.

> A30 Bodmin/Indian Queens (newbuild road): full briefing and early involvement of the main contractor enabled it to undertake joint management with the client of risks affecting the road scheme, by means of early archaeological investigations, participation in compulsory purchase procedures, agreement of an ecological strategy and resolution of access issues[244].

### 4.4.3 Risk sharing or joint risk management

One method of controlling the effects of risk is to reduce the main contractor's potential benefit should the risk arise. For example, PPC2000 provides that upon the occurrence of certain listed events of delay or disruption, the main contractor can recover certain categories of resultant cost but cannot recover additional profit or central office overheads or loss of profit on other projects. Thus, whatever the cause of the risk in question, the main contractor has a clear motivation to reduce its effects and is not tempted to translate a risk into financial gain[245].

An alternative risk management strategy is the sharing of risk, whether through joint venture arrangements where the client has a financial stake in the main contractor, or through adjustment of a building contract to provide that the main contractor and other parties share the cost effects of risk, for example in pre-agreed percentages. For example, on the Eden Project Phase 4 the client and the contractor and design team agreed to a target price under NEC2 by reference to which any savings of between £100,000 and £800,000 would be shared equally[246].

While any of these techniques may neutralise or reduce the potential for a main contractor or other team members to exploit risks to their advantage, they do not themselves achieve pre-emptive joint risk management actions that the contractor can undertake with other team members if appointed during the early phases of the project when it may still be possible to reduce the effects of the risks in question.

A 2001 ICE/DETR report specifically encouraged changes in working practices for the purpose of dealing effectively with ground risk and stated that 'Ground-related factors are a common cause of lengthy delays and large increases in building and construction costs. It is essential to put in place a risk management system to reduce and, if possible, avoid these problems and to exploit any opportunities for

---

[244] Project case study 7, Appendix A.
[245] PPC2000 Partnering Terms, clauses 18.5 and 18.6.
[246] See Project case study 11, Appendix A.

*Design, Pricing and Risk Management*

improvement that may arise'[247]. Their recommendations are equally applicable to management of other project risks:

> 'To provide more certainty of outcome in an increasingly fast-track and fragmented construction environment, the following are required:
> - good communication
> - a team approach to problem-solving
> - an integrated total project process
> - a risk-based approach to construction management and design'[248].

Ground risk offers a clear example of the way in which risk management can be incorporated into early project processes. Typically, the client and its advisers will have some knowledge of ground conditions and may commission certain additional information by way of investigations below ground prior to inviting main contractor bids. Such information in turn may be included in the tender documents. The successful bidder will have its own views as to the adequacy of the information provided and as to the risk involved in implementing the project without additional information. Early appointment permits the selected main contractor to express such views to the client and consultants and allows additional time for them together to examine:

- The main contractor's perception of the risk;
- Whether the client has additional information available to alter that perception;
- Whether additional site investigations would alter that perception;
- Whether early works packages (e.g. demolition) would alter that perception;
- Whether any aspect of the risk should be covered by insurance;
- Whether the client should assume all or part of the relevant risk;
- Whether any of these courses of action would give rise to the removal of, or significant reduction in, any amount of money allowed for such risk in the main contractor's price for the project.

Analysis of risk will only benefit the project if actions are undertaken that are based on the results of that analysis[249]. The above list is illustrative of the actions that can be agreed following joint analysis of risk at an early stage in the project, with time to take agreed risk management actions prior to agreement of a contract price and prior to start on site.

---

[247] ICE/DETR (2001), 20, 21.
[248] ICE/DETR (2001), 20, 21.
[249] See Smith *et al.* (2006), 34.

> Bermondsey Academy, Southwark (newbuild school): the site had restricted access, was next to a railway line and was overrun with Japanese knotweed. The preconstruction phase agreement required joint risk management which was led by the main contractor and through which the team identified risks, actions, costs and deadlines in a risk register. John Frankiewicz of Willmott Dixon stated that 'When commissioned at an early stage we will involve ... ground work contractors who may be able to identify potential risks that could be avoided through consideration in regard to orientation of the building or the location/availability of drainage services.' Preconstruction phase joint risk management also allowed the project team collectively to minimise the cost and time consequences of late changes in the location of temporary school facilities, caused by failure to obtain third party approvals[250].

### 4.4.4 Risk management and contracts

Commentators have long recognised the importance of risk management in the development of appropriate procurement strategies and contracts[251]. However, this does not address the question of whether risk management has a place in the contract itself. If construction contracts deal only with the construction phase of a project, it is difficult for them to govern any risk management procedures at all, and they do not offer a structure within which to place the series of stages of risk management that O'Reilly suggested are required to define agreed objectives and then 'to identify the potential courses of action for attaining these objectives'[252].

The conventional wisdom has been that risks cannot be reduced or eliminated through the creation of contracts, but that is arguably a position that only applies where the contract is created too late to influence risk management activities. Smith *et al.* stated that 'Risks cannot be eliminated through contracts but the strategy chosen for dealing with the risk can influence how they are managed and dictates how they are allocated'[253]. They proposed that:

> 'The contractual relationship is established during the tender period and during the early stages of post-contract award. If the project is to be a success, then it is during this phase that alignment must be

---

[250] Project case study 4, Appendix A.
[251] See, for example, O'Reilly (1995), 1.
[252] O'Reilly (1995), 1.
[253] Smith *et al.* (2006), 139.

achieved. This could take the form of a series of workshops where potential high risk sources are identified and joint plans are drawn up for dealing with them in the event that they materialise'[254].

These are practical proposals, but are unlikely to be fully committed to by the contractor if it is still working at risk, and will be undertaken too late if commenced only after creation of the construction phase building contract.

Working from the starting point of a conventional construction phase contract, Smith *et al.* recognised the need for significant change in order to achieve the integration of risk management with systems of project control and quality, and support the establishment of clear processes as a means to achieve this change. They stated that:

> 'The effectiveness of risk management is improved if all parties to a contract have the same appreciation of the identified risks. The contractor and the client should have similar views of the likelihood and potential effects of all risks. This can be achieved if pre-contract discussions between the client and the contractor ensure a clear mutual understanding of the relevant risks'[255].

Although they refer only to 'pre-contract discussions', logically the proposal by Smith *et al.* could be expanded to a joint contractual process undertaken during the preconstruction phase. Taking this a step further, it is arguable that risks can be eliminated, and certainly that they can be reduced, through the processes set out in a conditional preconstruction phase contract.

> Poole Hospital, Dorset (hospital refurbishment): preconstruction phase joint risk management processes, undertaken within agreed deadlines in accordance with a preconstruction phase agreement, made the main contractor aware of the need for a minimum number of operating theatres to remain available at all times while others were being refurbished and extended. The main contractor agreed to be flexible in its construction phase programme to allow for interruption in its work with minimal cost consequences, and the client and consultants identified additional non-urgent work on which the main contractor could redeploy its resources during such interruptions[256].

A preconstruction phase agreement can:

---

[254] Smith *et al.* (2006), 139.
[255] Smith *et al.* (2006), 94.
[256] Project case study 3, Appendix A.

- State any risk assumptions and requirements arising from the client's project brief or the main contractor's project proposals;
- Set out a timetable for agreed risk management actions to be undertaken during the preconstruction phase by the client, the main contractor, the consultants and certain subcontractors and suppliers;
- Describe a system for pricing the residual cost of risks that cannot be eliminated;
- Require, as pre-conditions to the project proceeding on site, agreement in respect of risk allocation, the costs of residual risks (within the client's budget) and all outstanding risk management actions.

The details of preconstruction risk management arrangements can be set out in a risk register, which should state clearly the party or parties responsible for particular risk management actions, the nature of those actions and the dates by which they need to be completed. A model form of risk register, taken from PPC2000, is set out in Appendix D.

### 4.4.5 Cost of joint risk management

Clients are likely to be concerned that joint risk management as part of a two-stage procurement process may allow a main contractor to inflate its priced risk allowances after securing an early appointment, so as to insulate its risk exposure at the expense of the client. In a properly managed process, the opposite should be true, as a cost-based approach to risk management should expose and eliminate any arbitrary percentage or lump sum contractor risk allowances that might otherwise be hidden in a single-stage bid.

Joint risk management activities may or may not lead to financial risk contingencies being reduced or eliminated sufficient to bring the project cost within budget. Even financial incentives do not ensure that project team members develop new risk management solutions or a greater willingness to absorb a particular risk. In the event that risk pricing remains higher than expected following early joint risk management, and if it adversely affects the affordability of the project, the team still have the opportunity to use other means such as value engineering to find cost savings that bring the project back within budget.

## 4.5 Preconstruction subcontractor appointments

### 4.5.1 Early subcontractor and supplier appointments

Early appointments of specialist subcontractors and suppliers can obtain not only additional design input but also improved risk management. Bennett & Jayes, in the *Seven Pillars of Partnering*, suggest early formation of a 'core team', which they suggest should include all those who contribute significantly to design decisions or to the management of the construction process[257]. They recognise that this should include key specialist contractors.

Bennett & Jayes envisaged this approach both at the 'Initial Creative Design Stage' and at the 'Plan and Control Stage'[258]. As to the tasks of this core team, Bennett & Jayes saw these as including the driving down of costs for the customer as well as maintaining or increasing profits for all firms involved. They suggest that the work of the core team should be driven by 'clear practical targets, ensuring suitable control systems are in place, monitoring progress, making sure problems are dealt with and targets are achieved'[259] all during the preconstruction phase of the project.

Early involvement of subcontractors and suppliers is regarded as fundamental to achieve the improved performance demanded by *Rethinking Construction*, which stated that:

'In our view, the supply chain is critical to driving innovation and to sustaining incremental and sustained improvement in performance' with the requirement for:

- Acquisition of new suppliers through value-based sourcing;
- Organisation and management of the supply chain to maximise innovation, learning and efficiency;
- Supplier development and measurement of suppliers' performance;
- Managing workload to match capacity and to incentivise suppliers to improve performance;
- Capturing suppliers' innovations in components and systems'[260].

The NAO in their 2005 report provided unequivocal government support in recommending early formation of an 'integrated project team', which they envisaged should comprise not only the client and consultants, but also 'specialist suppliers, including those involved in design'[261]. The NAO 2005 report suggested that early subcontractor and supplier involvement will 'maximise the opportunities for, and benefits of, value management and innovation'[262].

---

[257] Bennett & Jayes (1998), 74.
[258] Bennett & Jayes (1998), 74.
[259] Bennett & Jayes (1998), 74.
[260] Egan (1998), Section 45, 24.
[261] NAO (2005), 5.
[262] NAO (2005), 70.

*Early Contractor Involvement in Building Procurement*

The NAO recommended that a client working with an integrated project team formed at the earliest stages of a project will be 'better able to identify, articulate and share the objectives of the project'[263]. They pointed to school case studies which illustrate the need for a clear understanding by key suppliers that buildings need to be regarded as teaching and learning environments contributing to the wider community, and that all parties (including key suppliers), once able to share these objectives:

> 'Were better able to invest resources in identifying together the most cost-effective design solutions over the lives of the buildings; decide how the design and construction would impact on costs and health and safety during the construction and how operational efficiency could be maximised when completed'[264].

### 4.5.2 Barriers to early subcontractor and supplier appointments

Notwithstanding clear recommendations as to the benefit of early subcontractor and supplier appointments, there remains uncertainty as to the basis on which they should be appointed. Subcontractors themselves perceive that procurement strategies and forms of subcontract are a potential source of risk, and this view is presumably borne of bitter experience under hierarchical procurement and contract systems[265]. Greenwood refers to the Construction Industry Board's Code of Practice for subcontractor procurement, the principles of which require that such procurement should be 'principled, transparent and equitable', but points to the 1998 Constructors Liaison Group survey of subcontractors which concluded that the typical contractor/subcontractor relationship remains 'traditional, cost driven, and potentially adversarial'[266].

Changes to this entrenched position are necessary to get the best contributions from subcontractors and suppliers. Additional client and consultant influence can be brought to bear through establishment of a joint client/main contractor preconstruction process for early selection of specialist subcontractors and suppliers, particularly if this is agreed by reference to deadlines for its implementation in relation to each works package.

---

[263] NAO (2005), 70.
[264] NAO (2005) referencing Kingsmead Primary School and Blyth Community College, 70.
[265] See for example Greenwood (2001), 5.
[266] Greenwood (2001), 5.

*Design, Pricing and Risk Management*

### 4.5.3 Joint selection of subcontractors and suppliers

Joint client/main contractor selection of subcontractors and suppliers marks a departure from the traditional assumption that a subcontractor or supplier is either 'domestic' when selected by the main contractor or 'nominated' when selected by the client.

Domestic subcontractors and suppliers are selected by the main contractor as early as the tender stage or as late as the construction phase (subject to prior client approval)[267], with the main contractor solely responsible for their replacement in the event of default or insolvency and for meeting any consequent additional costs.

Nominated subcontractors and suppliers are selected by the client, usually prior to commencement of the construction phase, with the main contractor entitled to claim additional time and money in the event of subcontractor or supplier default or insolvency[268].

Joint selection of subcontractors and suppliers as part of a structured preconstruction phase process avoids the need for nomination and thereby also avoids consequent confusion as to the extent of the client's and the main contractor's liability for such subcontracts and suppliers. It permits direct client and consultant influence over which subcontractors and suppliers are most appropriate and offer best value, combined with acceptance of main contractor responsibility for their subsequent performance during the construction phase. Any sharing of risk for subcontractor or supplier default or insolvency can be agreed between the client and main contractor as part of the preconstruction phase process or later in response to particular circumstances.

> Bewick Court, Newcastle upon Tyne (housing refurbishment): the client and the main contractor jointly selected subcontractors and suppliers, including the cladding subcontractor, during the preconstruction phase. Responsibility for the performance of such subcontractors and suppliers following commencement of the construction phase rested solely with the main contractor, as it acknowledged when the cladding subcontractor went into administrative receivership. However, the client was persuaded to accept part of the cost and time consequences of this event in consideration of the main contractor's proposal to mitigate such consequences by itself assuming direct responsibility for the cladding package, using employees recruited from the former cladding specialist and materials purchased at a discount from the administrative receiver[269].

---

[267] See, for example, JCT 2005 SBC/Q, clause 3.7.
[268] See, for example, JCT 1998, clauses 35 and 36 and contrast the absence of equivalent provisions in JCT 2005 SBC/Q.
[269] Project case study 1, Appendix A.

A problem in early selection of subcontractors and suppliers as design team members is how best to obtain their design and risk management contributions while retaining a competitive process for their appointment to implement the works package they have designed. Three options emerge that will be more or less appropriate according to the nature of the works package, the length of the preconstruction phase and the commercial preference of the clients, consultants and main contractor:

- Selection of the subcontractor or supplier on the basis of it pricing fully developed consultant designs so that all of the subcontractor's or supplier's design, supply and construction activities are contained in a single price and appointment. This may not encourage bidding subcontractors or suppliers to offer design or other added value contributions prior to selection, as at the point of selection they will not wish to challenge consultant designs in case this prejudices their chances of success.
- Selection of the subcontractor or supplier on the basis of it pricing consultant designs, but by a selection process which expressly involves comparison of bidders' design proposals, thereby expressly encouraging a challenge to consultant designs by prospective subcontractors or suppliers prior to selection and appointment.
- Selection of the subcontractor or supplier to provide design contributions only in return for a fee, leading to finalisation of combined consultant, subcontractor and supplier designs which are then used as the basis for a further competitive process to select a subcontractor or supplier to implement the works package according to those designs. Bidders for the works package can include the subcontractor or supplier who undertook the design work, provided that such design does not favour its own proprietary systems.

A further question for the client and main contractor to address is what are the most appropriate criteria that should govern selection of subcontractors and suppliers. Aside from the price components that the client is seeking to fix, the criteria for selection of subcontractors and suppliers should be consistent with those used for selection of main contractors[270]. Otherwise, the temptation of clients and main contractors to focus primarily on subcontractor or supplier prices over and above other considerations may lead to an imbalanced team where selection of subcontractor and supplier members has not reflected the client's wider project priorities. For example, the Arup report to OGC comments in respect of PPC2000 that:

---

[270] See Chapter 4, Section 4.5 (Two-stage pricing) regarding the criteria recommended by CIRIA.

*Design, Pricing and Risk Management*

'The management provisions are dependent on sound information rising up through the supply chain and partnering terms with compatible conditions passing down to the Specialists who are subcontracted to the Constructor. The contract requires that these Specialist terms and conditions are to be compatible with the Partnering Terms and their management processes'[271].

Finally, the client and main contractor need to agree whether subcontractors and suppliers should be paid for their preconstruction phase contributions to the project. For the client to obtain the greatest value from the design and other contributions of subcontractors and suppliers, a commercial incentive of some kind is likely to be necessary. Bennett & Jayes specifically recommend that, as part of the 'core team' for a project, specialist contractors should be 'employed on a contractual basis that removes concerns about whether they will get paid … so as to empower all members to concentrate their efforts on doing their best work for the good of the project.'[272].

## 4.6 Perceived benefits of early contractor appointments

An organisation that has committed significant resource to the implementation of early contractor appointments is the Highways Agency[273]. The Highways Agency commissioned the Nichols report to review its major roads programme in March 2007, which recorded the following main potential advantages of early contractor appointments:

'Enables the contractor to influence planning decisions and design development at the most beneficial time:

- Potentially reduces preparation time for projects by 30–40%, by carrying out some parts of the development process simultaneously rather than consecutively;
- Gives [the client] access to detailed cost data to improve future estimates and output measures;
- Provides greater cost certainty, once the Target Price is agreed;
- Increases innovation which was being lost on D&B contracts; and facilitates value management and value engineering which can result in major cost and time savings;
- Provides the benefits from client and supplier working as a team;

---

[271] Arup 2008, 51.
[272] Bennett & Jayes (1998), 74.
[273] See Project Case Study 7, Appendix A and also the Highways Agency Procurement Strategy Review, Highways (2005).

- Enables tenderers, in the procurement phase, to prepare budget commentaries on the cost level which should lead to more accurate budget estimates;
- Requires the preparation of outturn estimates at key stages throughout the project cycle. This should lead to greater cost control during the construction phase, especially as the incentive formula is based on the Target Price, set before start of construction;
- Provides a team/alliancing spirit which leads to an open and honest process so that the real costs are highlighted early'[274].

The Nichols report also identified some potential disbenefits to early contractor appointments. For example, it noted:

- That the ECI process is still being refined after limited piloting;
- That contractors have successfully changed their culture and approach, but that the client needs further recruitment and training for this purpose;
- That there is some duplication of costs in early design, particularly where consultants are disconnected from the process and do not adopt the same team culture;
- That successful incentivisation requires robust cost estimates in creating a 'Target Price';
- That interests cannot be fully aligned while the client is inclined to take an overly optimistic view of costs and related risks and the contractor is anxious to cover its financial position.

These concerns illustrate a range of issues that in part can be addressed by agreeing clearer and more integrated preconstruction phase processes and in part through better education and training[275]. Nevertheless, the Nichols report recommended that the Highways Agency should continue with the use of early contractor appointments 'as the principal form of procurement for the present'[276].

## 4.7 Early contractor appointments and sustainability

The increasing attention paid by the Government and the construction industry to sustainability issues should further encourage the use of conditional preconstruction phase appointments for main contractors and specialist subcontractors. The need to maximise energy efficiency and to reduce waste highlight the significance of ideas that need to be developed not only by consultants but also by contractors and by their

---

[274] Nichols (2007), 32.
[275] See also Chapter 9, Section 9.5 (Education and training).
[276] Nichols (2007), 33.

subcontractors at every level. In order to evaluate and utilise main contractor and subcontractor ideas in response to sustainability initiatives, it is necessary to work with contractors and subcontractors during the planning and preconstruction phase of the project. Relevant contributions could include:

(1) Proposals as to the most buildable and least wasteful interpretation of the consultants' designs[277];
(2) Proposals in respect of training and employment[278];
(3) Proposals in respect of reduced waste and increased recycling[279];
(4) Proposals as to efficient use of energy on site, including modern methods of construction such as off-site fabrication[280];
(5) Proposals as to efficient use of energy by reduced maintenance and repair in the operation of the built facility[281];
(6) Creation of an acceptable Site Waste Management Plan[282].

> A30 Bodmin/Indian Queens (newbuild road): the early appointment of the contractor enabled it to schedule construction activities so that material was used to the maximum extent on site with very little going to landfill. In addition, the contractor was able to propose a number of environmental measures including reunifying the two halves of Goss Moor (a National Nature Reserve) previously divided by the old A30, by diverting the route of the new dual carriageway to the north and then degrading the old A30 back to its sub-base. While establishing informal agreements with land owners in advance of the required public inquiry, the contractor arranged for reptile fences to be put up early, so that the relocation of snakes and lizards could proceed in an orderly manner. This was commenced in early spring, as relocation is more difficult during the summer months, and was completed so as to avoid up to six months' slippage in the construction phase. Finally, the early appointment of the contractor enabled it to establish its supply chain arrangements early in the design process, so that such designs could be developed to suit locally available materials and locally available skills[283].

---

[277] See, for example, Project case studies 2, 4 and 6, Appendix A.
[278] See, for example, Project case study 9, Appendix A.
[279] See, for example, Project case studies 4 and 7, Appendix A.
[280] See, for example, Project case studies 7 and 8, Appendix A.
[281] See, for example, Project case study 4.
[282] See the Site Waste Management Plan Regulations 2008 (SI 2008/314) which require clients and contractors to create a Site Waste Management Plan and monitor minimisation of waste and appropriate waste disposal, and which assume that the main contractor will work with the client and its designers and planners at the design stage of the project.
[283] Project case study 7, Appendix A.

Although sustainability proposals can be assessed as part of a single-stage selection process, such a process depends on bidders putting forward their ideas on a speculative basis and integrating them with fixed price quotes. This would deny them the benefit of dialogue with the client and consultants that would allow them to formulate and submit better researched, more convincing proposals. The conflicting pressures of lowest price and added value through innovation are clearly evident in the arena of sustainability, with the result that bidders may hold back or compromise good ideas in order to reduce their bid prices. In a single-stage bid, there is also the increased risk that clients may reject proposals as unaffordable or unbuildable without the opportunity to investigate them in detail. By contrast, through joint working during the preconstruction phase of a project procured on a two-stage basis, the cost and quality benefits of sustainability proposals can be more thoroughly developed and assessed by a team that includes the main contractor.

This chapter has illustrated the potential benefits that can be obtained from early involvement of main contractors and specialist subcontractors in project designs, risk management and programming and from the early conditional appointment of the main contractor on the basis of agreed profit, overheads and agreed fees, so as to work with the client and consultants to control and agree remaining project costs.

# CHAPTER FIVE
# CLIENT LEADERSHIP, COMMUNICATION SYSTEMS AND BINDING PROGRAMMES

## 5.1 Introduction

The introduction and use of new preconstruction phase processes involving the main contractor and its subcontractors and suppliers will benefit from the use of all available techniques to ensure they are understood and put into effect. This chapter will examine potential for greater involvement of the client as a team member, and the importance of agreed communication systems and pre-agreed binding programmes to identify who does what by when during the preconstruction phase. Evidence from project case studies will illustrate the influence of these matters on project outcomes.

## 5.2 The role of the client

### 5.2.1 The need for client involvement

All projects have a client who sets the brief, appoints the other project team members (directly or indirectly) and makes the payments. Yet frequently the client delegates all project activities to other parties except for the statement of its initial requirements, the instruction of changes, the approval of designs and the expenditure of money. Should clients participate more closely in projects and can they do so without relieving other project team members of their responsibilities?

Latham proposed by reference to his recommendations for construction reform that the 'role of Government as client, along with leading private sector clients and firms, is crucial if the objectives of this Review are to be met'[284]. This led Latham to recommend that 'Government should commit itself to being a best practice client' and that 'A Construction Clients' Forum should be created to represent private sector clients'[285]. Thirty years earlier Banwell identified the importance

---

[284] Latham (1994), Item 1.17, 5.
[285] Latham (1994), Item 1.14, 5.

of the client in improving relationships between the members of the design and construction teams. Banwell's report identified a wide range of common problems in those relationships 'from the client (who must not be regarded as being outside the team) through his advisers, to the contractor and the contractor's man on the site'[286]. He recommended that 'New relationships are essential if the kind of advice which is needed for modern building is to be made readily available'[287]. It is argued that new client relationships need to be built up with all key parties, including the contractor, at an early stage in the life of the project.

The client is not always the end user, and frequently will have its own responsibilities to other parties for successful delivery of the project. If two-stage procurement and project management processes allow for a closer involvement of such a client in the project processes, this would also be a means to assist it in fulfilling its responsibilities to end users and to other stakeholders such as funders and regulators.

> Macclesfield Station (rail refurbishment): the client as a train operating company was responsible for obtaining consents from the owner of the site and the regulator of its business and for working within constraints imposed by these parties. These responsibilities were factors that led the client to choose and implement a two-stage procurement strategy by which the team established a fully designed, priced and programmed project prior to start on site, using a preconstruction phase agreement to define the required activities. This approach enabled the client to satisfy all its third party requirements[288].

Are most clients taking on the role that is required of them? A 1975 NEDO Report stated that 'The client has an important role to play in the construction process. Our case studies reveal widespread and conspicuous failure among public sector clients to give due regard to this'[289]. *Rethinking Construction* in 1998 also highlighted a lack of involvement by the construction client as a missing link in the project team. It recognised that 'Clients need better value from their projects', but also that the 'direction and impetus must come from clients'[290].

---

[286] Banwell (1964), 5, Section 2.8.
[287] Banwell (1964), 5, Section 2.8.
[288] Project case study 5, Appendix A.
[289] NEDO (1975), Section 7.11, 70.
[290] Egan (1998), Section 14, 13.

*Leadership, Communication and Programmes*

It would appear that, although it is in the client's own interests to take an active role, little had changed in the 23 years between these reports.

If the client participates in a project only by proxy through a consultant project manager, then the project manager becomes the only medium for a range of important decisions and recommendations, with the consequent risk of messages being lost in translation. While more direct client involvement is therefore desirable, there are circumstances where this is subject to necessary and understandable limits, for example because:

- Not all clients are professional clients: there is also the one-off client;
- Not all clients have time to acquire the knowledge necessary to break down barriers with the other project team members;
- Unless the client is its own expert, it has to rely on the expertise of others or engage a further set of experts to check the work of the first set;
- The client does not wish to risk relieving other project team members of their responsibilities by getting too involved.

Other reasons for the client holding back from closer involvement in a project may be less excusable, for example:

- The client may not wish to be available to deal with consultants and contractors if this requires it to consider complaints or requests that may highlight issues contrary to its interests;
- The client may prefer not to establish a clear understanding of pre-construction phase processes, but instead try to hold other project team members to their non-contractual, possibly over-optimistic promises.

For all of the above reasons, clarification of the exact parameters of client participation in a project, particularly during the preconstruction phase, is an important aspect of selecting the most appropriate procurement method, so as to be sure that it corresponds to available client resources and commercial client priorities[291].

Clients may be particularly concerned that closer personal involvement will lead them to diluting or confusing the roles and responsibilities of other project team members. These concerns can be addressed if the nature of the client's role and its interfaces with other team members are clearly described and limited as recommended by NEDO:

---

[291] See also Chapter 5, Sections 5.2.3 (Client involvement in preconstruction phase processes) regarding the scope of early client involvement.

- 'To act as a focal point' to integrate its interests;
- In conjunction with others as appropriate 'to define the scope and objectives of the project ... and to agree upon the methods of proceeding';
- 'To create a clear brief for the designers' and assist in its development;
- 'To react swiftly in obtaining any necessary strategic client decisions required during the currency of the design or construction phases';
- 'To monitor the overall progress and performance on the project'[292].

For the client to work as a member of the project team requires the client to engage with all other team members, namely with contractors and specialist subcontractors as well as with consultants. It also requires the client to agree and meet its own deadlines for decisions and other responses, and when appropriate to attend meetings in person.

Close client involvement in a project is supported by the principle that knowledge is power. However, one criterion for determining the appropriate level of client involvement is the question 'How much honesty can the client take?' Closer proximity to the project processes will make it harder for the client to deny the facts of a situation, including those where the client's expectations are compromised or deflated by circumstances outside the other project team members' control. This makes it harder for the client to cling to its original expectations or to postpone difficult decisions or ignore the problems confronting other project team members.

### 5.2.2 The client role under standard form building contracts

Most standard form building contracts provide for a specific individual or organisation to take on the role of 'project manager'[293] or 'architect/contract administrator'[294], 'client representative'[295] and to fulfil a number of functions including the issue of instructions to the main contractor[296]. Generally, the relevant individual or organisation has full authority to represent the client subject to stated limitations[297].

---

[292] NEDO Report 1975, Section 3.7, 26.
[293] Under NEC3 and GC/Works/1.
[294] Under JCT 2005.
[295] Under PPC2000.
[296] See, for example, JCT 2005 SBC/Q clause 3.10, NEC3 core clause 29.1, GC/Works/1, clause 40(1) of the Conditions of Contract and PPC2000 Partnering Terms, clause 5.3 and Perform 21 PSPC3, clause 7.1. The role of the project manager is considered further in Chapter 8, Section 8.2 (Preconstruction agreements and project management).
[297] See, for example, GC/Works/1 clause 4(1) of the Conditions of Contract and PPC2000 Partnering Terms, clause 5.2.

However, the above systems do not provide for direct involvement by an individual representative of the client itself (unless of course that individual is given the status of project manager or its equivalent). In fact, the delegated project management functions in most standard form construction contracts leave the client with very little direct contractual authority[298].

The omission of the client from most operational processes under a building contract is logical. For example, the main contractor cannot accept instructions from more than one party and there is no clear delegated authority if that authority could be countermanded or undermined by the client. However, lack of provision for the client to participate in important project processes can mean that in practice a client does not attend project meetings or read project documents until asked to make a payment or to grant an approval.

To omit clients from a contractual role in project processes deprives them of the ability to lead implementation of change as envisaged by *Rethinking Construction*[299] and of the ability to work with the construction industry in order to break down the barriers between the client and other project team members. Therefore, it is suggested that a procurement strategy and its contracts needs to ensure that the client has sufficient involvement to make decisions at appropriate stages in the project and is provided with the information it needs to make such decisions.

> Bewick Court, Newcastle upon Tyne (housing refurbishment): the build-up of open-book prices through joint selection of subcontractors and suppliers ensured that the client had full information regarding the cost of appointing a suitable cladding subcontractor and was aware of the difficulties of appointing an alternative such subcontractor. In addition, the client had agreed not to delegate its attendance at 'core group' meetings. As a consequence, the client was obliged to participate in looking at possible solutions, and had the information it needed to do so, when called to a core group meeting to seek an agreed solution when the main contractor announced that the selected cladding subcontractor had gone into administrative receivership[300].

In a set of two-party contracts, the client as the only common contracting party is the route via which one project team member obtains a

---

[298] There are exceptions to this, such as the right of the client under PPC2000 to confirm an instruction issued by its client representative before taking action against the main contractor for failing to comply with such instruction, PPC2000 Partnering Terms, clause 5.5.
[299] Egan (1998), Section 81.
[300] Project case study 1, Appendix A.

contractual remedy in respect of another project team member's failings. Where the client is entering into a series of separate contracts relating to the same project, there will be no contractual link between, for example, the architect, the structural engineer, the services engineer, the quantity surveyor and the main contractor. Each will have contractual rights and obligations only via the client as regards exercise of authority over each other and the effect of their communications with each other.

Exceptions to this restriction arise in the following (the first three of which are considered further in Appendix B):

- PPC2000, which as a multi-party contract creates a single system of delegated authority and communications enforceable by each team member against all of the others;
- The multi-party Partnering Agreement forming part of the Perform 21 suite of contracts (Perform 21 PSPCP) which creates communications procedures applicable to separate two party consultant appointments and building contracts;
- The multi-party JCT CE Project Team Agreement which is binding only in relation to sharing of risk and reward;
- The use of collateral warranties or third party rights provisions to establish direct contractual links between team members, for example, pursuant to the Contracts (Rights of Third Parties) Act 1999.

Two-party contracts leave the client as the focal point in all contractual disputes, attempting to assess the competing interests of project team members under their respective two-party contracts. The client needs some level of direct participation in the project to be able to fulfil this difficult function, and the absence of contractual opportunities for such direct participation may leave the client vulnerable in trying to untangle conflicting stories. For example, PPC2000 places considerable emphasis on client participation, primarily through its membership of the 'core group' (see also Section 5.3.4, later in this chapter, regarding the role of the core group). It provides that the client cannot delegate to the project manager its membership of the core group, and therefore has to participate directly with other team members in project processes such as the approval of designs and build-up of the supply chain[301].

### 5.2.3 Client involvement in preconstruction phase processes

Client involvement is particularly important in the early stages of the project when there is time for the client, in conjunction with all other

---

[301] PPC2000 Partnering Terms, clauses 5.2, 8 and 10.

team members, to participate in finalisation of requirements and the planning of project processes. N.J. Smith stated that 'It is paramount that all stakeholders (investors, end-users and others with a real interest in the project outcome, such as the project team, owner, constructors, designers, specialist suppliers) must be involved in the process, especially during the VP (Value Planning) and VE (Value Engineering) stages'[302]. As a starting point, the client should have a close working relationship with its design consultants during the preconstruction phase of a project. Yet no such client involvement is stated in the RIBA appointment other than by way of rights of approval at each stage of the architect's services[303].

A more formal system may be desirable in order to clarify the client's role during the early project stages, not only working with the design consultants but also with the main contractor and possibly key subcontractors and suppliers. Participation in preconstruction phase processes offers the opportunity for the client to establish appropriate links between other team members, to clarify its requirements and expectations, and to receive and consider proposals in respect of ideas that the client and its consultants may not have considered when drawing up the original brief.

It is also important to recognise the client's project responsibility. This should extend to responsibility for defining the parameters of the project, obtaining finance for the project, making key decisions during the course of design development and providing prompt approval and guidance to other team members. If a project is to be successful, the client needs to work with the other team members. Yet commentators have observed that through the proliferation of claims and disputes, as well as the development of fragmented team structures, clients and contractors have become increasingly removed from each other, and as a consequence project costs have increased and the construction industry has suffered[304].

> Project X (newbuild housing): the failure of the client to appoint a senior person to represent it on the project team led to a lack of client involvement in the project processes. Other parties adopted poorly organised and documented practices, contrary to those set out in the preconstruction phase agreement, which were not noticed by the client until they had given rise to contractor time and money claims[305].

---

[302] Smith N.J. (2002), 22.
[303] RIBA (2004), Services Supplement: Design and Management.
[304] See Smith *et al.* (2006), 136, as to client responsibilities and the risks of clients and contractors being increasingly removed from each other.
[305] Project case study 8, Appendix A.

## 5.3 The role of communication systems

### 5.3.1 Communication between organisations and individuals

Whatever the relationship established between organisations under forms of contract, Lock commented that 'When considering the project objectives it is easy but dangerous to forget that no objective can be achieved without people'[306]. MacNeil, when looking at trust established between individuals, noted the evidence 'that one of the most important of human techniques for developing trust is to make gifts … as proof that the giver is willing not to maximize utility from each exchange as such, a representation that he takes into account the interests of the other'[307]. This is also important to the success of building contracts, particularly in hybrid relational/neo-classical contracts governing conditional preconstruction phase processes during which personal and corporate relationships can be strengthened by evidence of early activities undertaken for reduced or deferred consideration.

To allow individuals to establish trust through provision of works or services for reduced or deferred consideration requires authority from the organisations to whom those individuals are accountable. For this to be reconciled with clear contractual arrangements will require terms of reference within which those individuals operate and interact with each other, as well as recognition as to the extent to which the compromising of their commercial interests, through the agreement not to maximise utility from each exchange, is voluntary rather than obligatory.

Such personal relationships therefore need the support of clear contractual communications systems so as to function efficiently. Agreed continuity of such relationships, agreed levels of delegated authority and agreed ways of reacting to unexpected events are among the tests of whether communication systems offer the means to avoid conflicts. MacNeil noted that 'Relational response to the breakdown of cooperation [thus] tends to be defined in terms of what is necessary or desirable to restore present and future cooperation'[308]. He referred to 'negotiation', which clearly depends on personal relationships, but also, surprisingly, referred to 'mediation' and 'arbitration' as other 'processes fostering cooperation'[309]. As mediation and arbitration require involvement of a third party, it is argued that they signal the failure of relationships and processes to provide a solution and that it is preferable to

---

[306] Lock (2000), 11.
[307] MacNeil (1981), 1047/1048.
[308] MacNeil (1974), 741.
[309] MacNeil (1974), 741.

*Leadership, Communication and Programmes*

provide a contractual means to maintain the direct engagement of the parties in finding a solution.

Most standard form building contracts provide for service of notices between organisations, but do not establish delegated authority and terms of reference for communication between individuals[310]. However, in view of the large number of people involved in delivery of a construction project, it is effective communication between individuals as much as between organisations that needs to be clearly established and understood.

The OGC include among their 'Critical Factors for Success' in their AEC *Construction Projects Pocketbook* 'Roles and responsibilities clearly understood by everyone involved in the project, with clear communication lines'[311]. There are also communication risks to be managed, such as:

- Too much communication, leading to a waste of time;
- Too little communication, leading to misunderstandings or missed opportunities.

N.J. Smith suggests that 'It is important that an explicit communication strategy is developed, and the necessary channels between the project participants are established. Rules for the use of these channels must also be put in place'[312]. Although project execution plans and organisational structures for teams are created, there is frequently no direct link between these documents and the contractual authority of particular individuals.

In broad terms, communications in relation to a construction project can be subdivided into two categories, namely the service of notices by one party to another and the attendance of meetings between the parties. It will be suggested that each can be improved through techniques such as the use of a 'core group' or 'early warning system' established early in the preconstruction phase of the project.

### 5.3.2 Notices and meetings

The medium for service of notices needs to be clearly agreed. Burke observed that 'The use of written communication should be encouraged because it addresses misinterpretation and forgetfulness. All important agreements and instructions should be confirmed in

---

[310] For example, JCT 2005 SBC/Q, clause 1.7 regarding giving or service of notices and other documents.
[311] OGC (2007), Construction Projects Pocketbook, 1.
[312] Smith N.J. (2002), 247.

writing'[313]. It is surprising that some standard form construction contracts permit an instruction to be issued verbally and then provide for written confirmation within a number of days. For example, GC/Works/1 permits some instructions of the project manager to be given orally, including resolution of discrepancies in the contract documents, removal and/or re-execution of work by the main contractor, suspension of work and execution of emergency work[314]. This seems likely to expose project team members to the serious risk of misunderstanding regarding the content of an oral instruction if the written confirmation of that instruction varies from what was understood verbally and acted on during the seven-day interim period.

Clear notice provisions are of particular importance to the client. In practice, the client is unlikely to be party to every notice served between other team members during the life of the project. However, the client is party to every contract that it awards in relation to the project and will be the recipient of any formal notices served under those contracts. It is, therefore, important for the client to establish in its building contract and consultant appointments a clear and consistent approach to communications that establishes:

- Notices to be sent and received by the client;
- Clarity as to delegated authority of other parties to send and receive notices on behalf of the client;
- Agreed media for written communications with appropriate evidence of receipt.

As regards evidence of receipt of notices issued between team members, Lock commented 'For every instruction which is sent out (on a project), a resulting feedback signal must be generated. Otherwise there will be no way of knowing when corrective actions are needed'[315]. Bennett made a similar comment that 'Feedback is absolutely crucial for construction to achieve improvements in its performance. It operates at every level and all teams should use systematic feedback to control their performance'[316]. A clearly defined system of communication establishes the discipline necessary to ensure that the agreed media are used and the required acknowledgement of feedback is obtained.

Turning to meetings, many of these are required at every stage in a construction project. However, if meetings are not clearly structured

---

[313] Burke (2002), 247.
[314] For example, GC/Works/1 Two Stage Design & Build, clause 40(3) which provides that 'Oral Instructions shall be immediately effective in accordance with their terms, but shall be confirmed in writing by the PM within 7 Days'.
[315] Lock (2000), 482.
[316] Bennett (2000), 187.

as to their attendees, timing and purpose, there is the risk that they can be poorly managed, time consuming and wasteful. Project managers have noted that meetings themselves can result in presentation of excuses by participants as to why they have not carried out the actions requested of them. At their worst, Lock expressed the view that meetings start with explanations of delay or inefficiency and end with promises as to how matters will be dealt with differently – only to lead to a further round of excuses at the next meeting[317]. This concern supports the importance of connecting meetings with the agreement and implementation of programmes specifying the agreed activities of each team member so as to leave the minimum room for doubt or ambiguity.

As Burke observed 'An effective way to achieve commitment is to make the person aware of the cost of any delay to the project'[318]. This awareness can best be created at meetings if they are used to track progress against agreed briefs, costs and programmes by reference to which the risks and consequences of prospective delays or other failures can be more easily identified and communicated. In order that project meetings are run efficiently and the individuals attending meetings are confident as to how they should behave and do not feel that their authority may be challenged, it is important that the structure, terms of reference and organisation of those meetings are clearly understood.

A further problem can arise where team members believe they are spending time at meetings for which they have not made any financial allowance when costing the project[319]. A report by Barlow *et al.* in relation to partnering identified concerns that partnering had led to a disproportionate increase in the amount of time spent by the parties in communicating with each other, through an excessive number of points of contact involving more senior staff than would normally be appropriate. One of the interviewees of Barlow *et al.*, a specialist supplier, was concerned about spending time in meetings that were not relevant to his trade because a consultant thought it necessary that all meetings should involve everyone engaged on the project[320]. This may be attributable to the tendency, when implementing new approaches to procurement, to over-resource meetings in order to understand and influence the process of change. The risk is that the level of resource becomes uneconomical – which leads to increasing delegation to more junior staff or failure to attend meetings. This can have a debilitating effect on the morale of team members and needs to be avoided.

---

[317] See Lock (2000), 510.
[318] Burke (2002), 200.
[319] See also Chapter 9, Section 9.3.2 (Cost and time to create agreements).
[320] Barlow *et al.* (1997), 55.

A communications system can set out agreement between the team members as to who should attend what meetings, when and for what purposes. While it may not be possible to prohibit parties from calling unnecessary meetings, an agreement can clarify the system for calling meetings, can limit the number of participants and can avoid the distraction and delay of the participants having to spend time working out their own terms of reference.

### 5.3.3 Creation of a contractual core group

The joint establishment of an agreed system of communications should itself be an early preconstruction phase process, and is of particular importance in securing the effective management of other preconstruction phase processes by reference to a preconstruction phase programme. To the extent that such processes may involve the main contractor and key subcontractors and suppliers in new ways, communications systems agreed with these parties will be necessary in order to ensure that the timing and nature of their contributions are fully understood and are properly integrated with the roles and responsibilities of other team members.

The establishment of a 'core group' of key individuals representing project team members is an approach recognised in certain standard form building contracts[321]. The members of a core group need clarity as to:

- Their levels of delegated authority;
- Their terms of reference;
- The circumstances in which they meet;
- The procedure governing their meetings;
- The means by which they reach decisions;
- The limits on replacement or substitute members.

> Bermondsey Academy, Southwark (newbuild school): although the main contractor and design consultants had bid for the project jointly, there were misunderstandings as to the interfaces between their respective design responsibilities. These were resolved during the preconstruction phase, using core group meetings to ensure open discussion and to maintain progress in design development while negotiations were completed[322].

---

[321] NEC3, PPC2000 and Perform 21 all recognise the role of a core group. See Appendix B.
[322] Project case study 4, Appendix A.

A core group needs to be properly established, and its members should be obliged to meet even when they do not want to, for example whenever there is a potential dispute. With clarity and discipline as to the structure of its meetings, a core group can resolve apparently intractable problems.

> Project X (newbuild housing): despite failings by the client, the architect, the engineer and the main contractor that none of them wished to acknowledge, a series of core group meetings were attended by all of them in accordance with the contractual system originating in their preconstruction agreement. This system provided a basis for the parties to recognise their respective shortcomings and to inch away from entrenched positions until they achieved a compromise. The main contractor revealed that its core group member had brought a notice of adjudication to serve at one of the core group meetings, but that in the light of the direction of discussions towards a settlement he had kept it in his pocket instead[323]. All claims and disputes were settled by the core group, guided by the partnering adviser[324], without the use of formal dispute resolution procedures.

Joint decision making processes through a medium such as a core group offer a technique to deal with the neo-classical features of preconstruction phase agreements identified by MacNeil, namely 'gaps in their planning'[325]. Although MacNeil envisages third party assistance to fill these gaps[326], agreement between the core group as representatives of the team members is more likely to be accepted by team members as a means of developing and completing the details of their contractual relationships. The core group is therefore a forum for overcoming the conditionality of the preconstruction phase agreement so as to achieve consensus on a developed brief, proposals, prices and programme sufficient for the parties to commit unconditionally to the construction phase of the project.

The core group needs the support of the contract not only in clarifying its terms of reference, but also in guiding it as to the positions established between the parties at each stage in the project as they move from incomplete to complete cost, time and quality information. The conditional preconstruction phase agreement should map out the iterative development of complete information, and the methodology

---

[323] Project case study 8, Appendix A.
[324] See Chapter 9, Section 9.6 (The role of the Partnering Adviser).
[325] MacNeil (1978), 865.
[326] See Chapter 2, Section 2.2 (Recognised categories of contract).

and agreed parameters of that iterative development must be clear so as to support the core group in maintaining consensus while the remaining details are completed.

The core group also fulfils a valuable problem-solving and dispute resolution function in resolving differences that emerge between members of the supply chain. Core group members can only work with the information set out in or developed pursuant to the contract and will find it difficult to maintain consensus if called upon to reach subjective decisions or to make assumptions as to the parties' roles and responsibilities. However, if the right individuals are chosen, and if the contract terms and the machinery governing their terms of reference are clear, the problem-solving work of the core group can save the client and other team members a great deal of time and money.

The Arup report to OGC observed in its commentary on PPC2000:

> 'The creation of a Core Group to guide the project also has a dispute resolution function. This ensures the visibility of problems and any impact of those problems upon the project irrespective of the point in the supply chain at which they are found ... In providing these processes it is expected that the parties will find that the terms of the contract provide a swifter and more cost-effective way of resolving points of difference than they might obtain from other dispute resolution mechanisms available such as adjudication or litigation'[327].

### 5.3.4 Contractual duty to warn of problems

In addition to organising channels for communication, it is also important to remove barriers to communication, particularly as regards notification of problems. If a communication system is not successful in allowing the parties to alert each other as to problems, it is likely that such a system will become a basis merely for keeping records after the event and will be of little value to the team. Warnings need to be issued as soon as a problem arises, and need to be issued to the correct party on the understanding that notification will lead to timely decisions and actions.

Records created after the event are more likely to be used to allocate blame than to initiate actions, and therefore will not help the project team in resolving problems or mitigating their effects. N.J. Smith stated that:

> 'A system of communication needs to be planned and monitored, otherwise information comes too late, or goes to the wrong place for

---

[327] Arup 2008, 38 and 39.

decisions to be made. The information then becomes a mere record, and is of little value. The records are then used to allocate blame for problems, rather than to stimulate decisions which will control the problems'[328].

A communication system can only support efficient teamwork if it encourages early notification of problems that could otherwise degenerate into disputes, so as to reduce the risk of allocation of blame after the event. The courts have increasingly taken the view that contractors as well as design consultants and professional advisers have a duty to warn their clients of any design defects of which they become aware[329]. A contractual communication system can clarify and extend such a duty to warn to apply to any potential problem, linked to the role of the core group as the forum at which to review the problem when notified.

Lack of contractual clarity as to whether project team members have a duty to warn each other of actual or potential problems will naturally lead parties to err on the side of caution. However, Bennett observed that:

> 'When a key target is in danger of being missed, this must be treated as a crisis and clear, effective action taken quickly to get the work back on its planned course. A control system which is not used to provide this steady, systematic control is simply a waste of resources. Once it is discredited by being ignored, nobody will bother to provide accurate or up-to-date feedback'[330].

PPC2000, for example, provides that if a client or consultant deadline is missed, then the contractor must give early warning to the client not more than five working days after the expiry of the relevant time limit[331].

It can be argued that parties will always be reluctant to expose themselves to increased liability through early warning, even if that early warning is reviewed by a core group, as there is no guarantee that the core group will reach a conclusion other than to allow the parties to enforce their respective contractual rights. However, there are circumstances where:

- The contractual rights of the parties are a solid starting point that enables them to consider whether an alternative response outside the contract is appropriate in particular circumstances.

---

[328] Smith N.J. (2002), 12.
[329] As established in, for example, *Tesco Stores Ltd* v. *The Norman Hitchcox Partnership Ltd & Others* (1997), 56 CON L.R. 42.
[330] Bennett (2000), 186.
[331] PPC2000 Partnering Terms, clause 18.3(i).

- The contract does not cover every eventuality and, given the opportunity, the parties can put forward intelligent proposals to vary the contractual position if these offer a better solution.
- Such proposals can be acceptable to the other parties if they serve their respective interests better than enforcement of strict contractual rights.

> Bewick Court, Newcastle upon Tyne (housing refurbishment): the establishment of a core group, with an agreement to meet and consider early warning given by any team member, provided a forum to which the project manager could notify his concerns when the cladding subcontractor went into administrative receivership. This allowed the main contractor to present an innovative solution for approval by the client that did not adhere strictly to contractual terms but that reduced potential delays, involved very little additional cost to the client and preserved a reasonable level of main contractor profit. A further early warning by the project manager later in the project enabled the core group to deal with suspension of work due to third party interference, again agreeing a solution outside the contract terms at minimum cost to the client[332].

The successful use of early warning depends on the parties overcoming their instinctive wish to remain silent rather than be implicated by notifying problems apparent in another party's performance or invite trouble by notifying problems in their own performance.

Eggleston observed in relation to early warning under NEC3 that the relevant core clause 16.1 'is clearly more than a mechanism for one party informing the other of its (the other's) faults. It requires confession of the parties' own faults'[333]. It is, however, questionable whether the requirement for notification of a party's own faults or those of another party to the project manager under NEC3 is more or less of an inducement to overcome instinctive reticence than the requirement of notification to the core group under PPC2000. The project manager under NEC3 is accountable only to the client[334], whereas the core group under PPC2000 is representative of all team members and has a duty to seek solutions to potential differences and disputes[335].

---

[332] Project case study 1, Appendix A.
[333] Eggleston (2006), 117.
[334] See also Chapter 8, Section 8.2.3 (The role of the project manager in integrating other team members).
[335] PPC2000 Partnering Terms, clause 27.2.

It is not possible to define rigidly in a contract all the circumstances in which early warning should be given. There is therefore a risk that, if the parties are entitled to notify problems, they will use early warning excessively as a means to seek contractual waivers or leniency or simply to cause a distraction. However, at worst that risk is only the risk of wasted time and should be manageable by means of peer group pressure and common sense, particularly if it is clear that early warning does not oblige the parties to compromise their other contractual rights. Hence, I would argue that the risk of excessive early warnings is less damaging than the risks inherent in the parties hiding problems from one another. Eggleston noted in relation to NEC3 that 'Some degree of common sense and some tests of reasonableness and seriousness must be applied to avoid trivial matters obscuring the true purpose of the provisions'[336].

## 5.4 *The role of binding programmes*

### 5.4.1 Programming and project management

Programming is a key tool for planning a project. For all but the most simple projects to succeed, programming is at the heart of project planning and needs to be the subject of continuous monitoring and updating. Smith *et al.* stated:

'The activities of designers, manufacturers, suppliers, contractors and all other resources must be organised and integrated to meet the objectives set by the client and/or the contractor. In most cases, the programme will form a basis of the plan. Sequences of activities will be defined and linked on a timescale to ensure that priorities are identified and that efficient use is made of expensive and/or scarce resources'[337].

However, Smith *et al.* also observed that 'It is very difficult to enforce a plan which is conceived in isolation and it is therefore essential to involve the individuals and organisations responsible for the activities or operations as the plan is developed'[338].

However, programming is often left too late. Bennett noted that 'Time pressures on many projects mean that the set of coordinated method statements, programmes and budgets will still be under development after construction has started'[339]. Delay in creation of a

---

[336] Eggleston (2006), 117.
[337] Smith *et al.* (2006), 6.
[338] Smith *et al.* (2006), 6.
[339] Bennett (2000), 173.

programme until after start on site is widely accepted in the construction industry, even though such delay means that the contractor at the point of start on site is immediately under pressure to create an additional document that is necessary for it to work in an efficient manner, with the risk that adoption of this document will be further delayed if it is not accepted by the client or project manager or the contractor's own supply chain.

Failure to agree a programme will risk delays in key project activities and consequent losses to project team members. The absence of a programme also leaves the project manager without controls over who does what and in what sequence, and leaves the other project team members ignorant of the expected timing of their own contributions and those of other parties[340]. Bennett & Jayes suggested that all project processes should be 'planned as far ahead and in as much detail as possible without constraining the core team in its search for the best possible answers' and that it is important that 'Everyone in the project team fully understands the process they are working through and have bought into it'. While they stated the importance that 'Programme milestones are reliably met,' they also qualified this to the extent that 'where innovation is needed, milestones are interpreted flexibly on the basis of providing just sufficient information to avoid delaying the project'[341]. The NAO emphasised in their 2005 report the importance of improved programming and included among the characteristics of successful construction clients 'The creation of effective construction programmes'[342].

Banwell emphasised the importance of programmes in his observation that 'Insufficient regard is paid to the importance or value of time and its proper use in all aspects of a project, from the client's original decision to build, through the design stages and up to final completion'[343]. The need for integration of design and construction activities that runs through all Banwell's recommendations underlines the need for programming to cover not only the construction phase but also the preconstruction phase of a project. In Banwell's view it is the duty of those who advise a client 'to make it clear that time spent beforehand in settling the details of the work required and in preparing a timetable of operations, from the availability of the site to the occupation of the completed building, is essential if value for money is to be assured and disputes leading to claims avoided'[344].

---

[340] See also Chapter 8, Section 8.2.4 (The use of programmes by project managers).
[341] Bennett & Jayes (1998), 80.
[342] NAO (2005), 28.
[343] Banwell (1964), 3, Section 2.2.
[344] Banwell (1964), 3, Section 2.3.

## 5.4.2 Preconstruction phase programmes

A key function of a programme governing the preconstruction phase will be the completion of all activities that are preconditions to proceeding with the construction phase, including the finalisation of agreed designs, prices and supply chain arrangements, the establishment of an acceptable understanding regarding project risks and a variety of other matters such as satisfying health and safety requirements, obtaining third party consents, securing full project funding and agreeing the construction phase programme[345].

Smith *et al.* identified the following interfaces as critical to the creation of a successful programme, and emphasised that these interfaces need to be managed efficiently:

- Between different design consultants;
- Between design consultants and specialist subcontractors and suppliers;
- Between the design process and the procurement process;
- Between the design process and the construction process;
- Between the procurement process and the construction process;
- Between the project and other projects[346].

Most of the above are activities that need to be undertaken and programmed primarily during the preconstruction phase. The level of detail in preconstruction phase programmes will differ according to the complexity of the project.

> Bewick Court, Newcastle upon Tyne (housing refurbishment): the preconstruction phase programme set out activities over a three-month period ending with start on site, including design development, subcontractor selection and finalisation of prices. Each activity was stated to be the responsibility of one or more team members[347].

A successful preconstruction phase programme should identify the deadlines and responsible parties for each of the following activities:

- Design development submissions;
- Surveys and investigations;

---

[345] See sample preconstruction phase programmes annexed to Project case studies 1 and 7, Appendix A.
[346] See Smith *et al.* (2006), 59.
[347] Project case study 1, Appendix A.

- Cost plan submissions;
- Value engineering and value management reviews;
- Procurement processes for selection of subcontractors and suppliers;
- Pricing processes for all work and supply packages;
- Risk management actions (linked to any risk register);
- Client approvals and comments in response to each submission and proposal by other team members;
- Submission of applications for third party approvals;
- Funding, land acquisition and other client preconditions to commencement of work on site;
- Satisfaction of health and safety preconditions and other legal and regulatory preconditions to commencement of work on site;
- Satisfaction of insurance and security preconditions to commencement of work on site.

As regards each of the above, the preconstruction phase programme should state the relevant activity or requirement, the party or parties responsible and the period or deadline for the relevant activity. A sample preconstruction phase programme, comprising the form of 'Partnering Timetable' under PPC2000, is set out in Appendix E.

Certain activities in a preconstruction phase programme will need to be subdivided to ensure sufficient clarity. For example, in respect of subcontractor or supplier tenders, the activities and deadlines will need to be broken down into the following headings in respect of each work/supply package:

- Creation of drawings and specifications in sufficient detail for pricing – to be prepared by design consultants with input from the contractor;
- Agreement of invitation to tender and form of subcontract/supply contract – to be prepared by the contractor;
- Agreement of list of subcontractor/supplier tenderers – to be proposed by the contractor and other team members as appropriate;
- Issue of invitations to tender to prospective subcontractors and suppliers – usually by the contractor;
- Return of tenders from subcontractors/supplier bidders;
- Review of subcontractor/supplier tenders – by the contractor with the project manager and other team members as appropriate;
- Recommendation and client approval of preferred subcontractor/supplier prices and proposals.

In addition, there may be certain projects where the preconstruction phase is of a duration and complexity such that all preconstruction phase activities cannot be agreed from the outset. For example, it may be appropriate for the client and main contractor, with the other team members, to agree a series of activities necessary to establish the feasibility of the project or to obtain planning approval, and then (once such feasibility or approval is established) to agree the remaining activities through to start on site. Breaking the preconstruction phase agreement down into these stages is an option that the team should adopt only if justified by the circumstances, as the agreement of activities and deadlines in stages breaks the continuity of the team members' preconstruction phase commitments.

> A30 Bodmin/Indian Queens (newbuild road): the preconstruction phase programme contained 478 activities ending with start of construction. These activities included contractor contributions to design development and joint risk management (including preparation of an environmental statement) and each had a stated deadline. In view of the particular importance and complexity of the public inquiry process, a subset of this programme recorded 225 client, consultant and contractor activities and deadlines relating to preparation for and attendance at the public inquiry[348].

### 5.4.3 Programming consultant design outputs

The need for the detailed programming of design activities is underlined by the perceived difficulty in establishing clear interfaces between design team members and in establishing deadlines that do not damage or excessively constrain the creative design process. Lock observed that:

> 'In most engineering design offices and other software groups, highly qualified staff can be found whose creative talents are beyond question. But, while their technical or scientific approach to project tasks might be well motivated and capable of producing excellent results, there is always a danger that these creative souls will not fully appreciate the importance of keeping within time and cost limits'[349].

N.J. Smith stated that:

> 'The creative element of the designing process requires a period of synthesis that cannot always be "forced", the subconscious mind

---

[348] Project case study 7, Appendix A.
[349] Lock (2000), 479.

needs to work on the problem ... any designers resent the "imposition" of a mechanistic management regime, since they feel this constrains their ability to design effectively ... Since a fundamental aspect of design is the element of creativity, and the difficulty this brings in terms of an accurate estimate of the time needed to complete a design task, this must be considered in the management regime'[350].

Whatever the scope that must be allowed for design consultants to work creatively, the timing of their outputs at each stage of the design development process still needs to be agreed if the client and other team members are going to be able to plan the project. In any project, designs need to be sufficiently developed in order to be priced by the main contractor and its supply chain, and also to obtain required planning consents and other third party approvals. In a project structured so as to obtain early contractor involvement, designs need to be delivered to an agreed level of detail in time for contractor review and input, such review and input being activities which themselves need to be subject to agreed deadlines.

N.J. Smith recognised that:

'One of the fundamental requirements for effective management of design is an efficient flow of information between the participants in the project. This applies particularly to those that have an input to the design phases of the project, and a list of such participants would include at least:

- The promoter and appropriate groups within the promoter organisation (such as their design department);
- The users/operators of the project deliverable(s), which may or may not be part of the promoter organisation;
- The project manager;
- Team leaders in the project team;
- The design manager;
- The lead designers;
- The design team leaders;
- Sub-designers and appropriate team leaders within those groups;
- Design approval consultants acting on behalf of the promoter;
- Design checking consultants;
- Local authorities;
- Statutory bodies'[351].

---

[350] Smith N.J. (2002), 243, 244.
[351] Smith N.J. (2002), 246.

The RIBA and ACE standard forms of appointment allot substantial fees to preconstruction phase design activities, but do not provide for agreement of fixed deadlines for their completion. However, the progression of the preconstruction phase of a project depends substantially on design consultants meeting their deadlines in respect of each stage of design development.

The RIBA (2004) Appointment defined a 'Timetable' as 'The period of time which the Client wishes to allow for completion of the Services'[352] and related notes suggest that the parties should 'Identify any key dates that the Client wishes to achieve'[353]. Meanwhile, the ACE (2002) Agreements provided for 'Timeliness' whereby 'the Consultant shall use reasonable endeavours to perform the Services in accordance with any programme agreed with the Consultant from time to time'[354]. Both of these are very light obligations, when read alongside the provisions under most standard form construction contracts, whereby a delay caused by a consultant will give rise to a main contractor claim for additional time and money[355].

As observed by Nick Lane in the context of the 2012 Olympics 'Having established the programme's fundamental importance, will it cover design stages as well as construction? Here, we might ask whether consultants will be tied into time deadlines. Why shouldn't they? Timing of the design will be just as important as timing of the construction'[356].

> Watergate School, Lewisham (newbuild school): the project team members were committed to design excellence, but delays started to occur in the issue of designs for pricing by subcontractors and suppliers. The prior agreement of a preconstruction phase programme ensured that team members were reminded of binding deadlines for design deliverables, which in turn ensured the pricing of these designs in time to achieve the required cost certainty and avoid a delay in start on site[357].

---

[352] RIBA (2004), Conditions of Engagement Definitions, 9.
[353] RIBA (2004), Notes for Architects.
[354] ACE (2002), B2 Obligations of the Consultant, clause 2.9.
[355] For example, JCT 2005 SBC/Q, clauses 2.29.6 and 4.24.5 of the Conditions, GC/Works/1 clauses 36(2)(b) and 46(2)(a) and (c) of the Conditions of Contract, PPC2000 Partnering Terms, clause 18.3(i), and NEC3 core clause 60.1(5), (6).
[356] Lane (2005).
[357] Project case study 2, Appendix A.

### 5.4.4 Programming other preconstruction phase activities

Adherence to deadlines during the preconstruction phase is important in order to create sufficient time for subcontract tender documents to be developed and issued to prospective subcontractors and suppliers and for those subcontractors and suppliers to return not only prices but also qualitative proposals for evaluation. Once certain subcontractors and suppliers are in place, time may also need to be found in the preconstruction phase programme for them to contribute alongside other team members to value engineering exercises.

Programmes are important to achievement of effective risk management actions. It is also important that they recognise external influences that may cause delays outside the control of project team members and clarify where possible the impact of such external influences. R.J. Smith observed that 'The owner should consciously decide to make risk management an integral component of program/project planning and engineering as well as contract administration. This is most effectively done by a coordinated sequence of activities'[358].

If other preparatory activities need to be approved on or off site ahead of unconditional authority for the construction phase to proceed, for example mobilisation or long lead-in commitments to materials or equipment, then these to should be timetabled in the preconstruction phase programme so that the client can prepare for any required early expenditure.

> Macclesfield Station (rail refurbishment): the client needed to ensure that all issues were dealt with during the preconstruction phase. The programme for preconstruction activities identified the sequence of agreed periods and deadlines for each stage in design development, for the procurement of subcontract and supply packages, for agreed risk management exercises, for the impact of the grant of owner and regulator approvals, and for the finalisation of a price and construction phase programme[359].

### 5.4.5 Early agreement of construction phase programmes

In order to agree a construction phase programme prior to commencement on site, the parties need to agree the timing of their respective contributions to such a construction phase programme, with a further

---

[358] Smith R.J. (1995), 66.
[359] Project case study 5, Appendix A.

period for it to be signed off alongside all other preconditions to the construction phase of the project proceeding. Smith *et al.* suggested that construction programmes submitted by bidders can be a useful comparator for assessing their respective bids: 'The construction programme submitted by the contractor should be taken into account as part of the technical evaluation as it indicates the contractor's overall approach to the work, although many standard forms do not require a programme to be submitted at tender'[360].

Clients have been advised not to create a construction programme as a contractual document, as this would be likely in turn to give contractual status to the main contractor's method statements, any variation to which could give rise to greater time and/or cost entitlements for the main contractor. This was illustrated in the case of *Yorkshire Water Authority* v. *Sir Alfred McAlpine & Son (Northern) Limited* where an approved bar chart and method statement were signed as part of the contract and assisted the main contractor in its claim that it was entitled to follow that method statement or to seek a variation order with appropriate time and cost consequences[361].

It is arguable that to create a construction phase programme as a project management tool, but for the parties not to be contractually bound by it, leaves uncertainty in relation to matters where team members need to make clearly defined mutual commitments to meet specific deadlines. A contractual distinction therefore needs to be made between the main contractor's non-contractual method statements on the one hand and a contractual set of 'key dates'[362] on the other hand. The latter should include all of the following:

- Date or dates of possession of the site and each part of the site;
- Programmed interruptions or restrictions of possession of the site;
- Prior consultation requirements with users of the site;
- Preconditions or procedures prior to commencement on any aspect of the project;
- Time limits for completing outstanding designs of any element of the project;

---

[360] Smith *et al.* (2006), 162.

[361] *Yorkshire Water Authority* v. *Sir Alfred McAlpine & Son (Northern) Limited* (1985) 32BLR 115. However, see Chapter 5, Section 5.4.6 (Binding programmes and the SCL Protocol) as to binding programmes recommended by SCL Protocol, and see Appendix B, Section 6 regarding binding programmes under, for example, NEC3.

[362] JCT 2005 provides for key dates in respect of release of information to the main contractor pursuant to an 'Information Release Schedule', e.g. JCT SBC/Q clause 2.11. The NEC3 contracts each provide for adherence to 'Key Dates', e.g. NEC3 core clause 30.3. PPC2000 provides for key dates during the preconstruction phase to be set out in a 'Partnering Timetable' and during the construction phase in a 'Project Timetable', PPC2000 Partnering Terms, clause 6.

- Time limits for finalising details of provisional sum items, pricing those items, authorising their expenditure and placing the requisite orders in time for their implementation on site;
- Arrangements regarding division of the project into sections;
- Time limits for commissioning procedures and handover of the completed project or any part of it.

These are all matters where the main contractor does not act in isolation and is reliant on other team members fulfilling their obligations within specific time limits or on the activities of third parties over whom it exercises no control.

Lock suggested that, in order to schedule and control a project, 'network planning' should be sufficiently detailed to:

'Enable the following types of events to be identified, planned and monitored or measured:

(1) Work authorisation;
(2) Financial authorisations from the customer;
(3) Local authority planning application and consent;
(4) The start and finish of design ... if the duration of the design task is longer than two or three weeks, it might be advisable to define separate shorter activities corresponding to design phases;
(5) Release of completed drawings for production or construction;
(6) The start of purchasing activity for each sub-assembly or work package;
(7) Issue of invitations-to-tender or purchase enquiries;
(8) Receipt and analysis of suppliers' or subcontractors' bids;
(9) ... The issue of a purchase order to a supplier or subcontractor;
(10) Material deliveries;
(11) The starts and completions of manufacturing stages;
(12) The starts and finishes of construction subcontracts, and important intermediate events in such subcontracts;
(13) Handover events of completed work packages'[363].

> Poole Hospital, Dorset (hospital refurbishment): during the preconstruction period, the team agreed a contractually binding construction phase programme that provided dates and periods of time to reflect:

---

[363] Lock (2000), 216.

> - Restrictions on access to the site through the interior of a functioning hospital;
> - Rotation of work in operating theatres so that a minimum number remained available at any time;
> - The time required to switch resources to alternative work in the event of interruption to work in the operating theatres[364].

## 5.4.6 Binding programmes and the SCL Protocol

The Society of Construction Law Delay and Disruption Protocol (the 'SCL Protocol')[365] was launched in October 2002 with the express aim of encouraging project team members to avoid costly disputes by tackling issues early on through pre-planning, risk management and transparency. It recommends that:

- 'The Contractor should prepare and the Contract Administrator (CA) should accept a properly prepared programme showing the manner and sequence in which the Contractor plans to carry out the works;
- The programme should be updated to record actual progress and any extensions of time (EOTs) granted;
- Contracting parties should also reach a clear agreement on the type of records that should be kept'[366].

Although the SCL Protocol is intended to encourage best practice, it leaves some surprising omissions, for example:

- An effective programme should govern the activities of all project team members, yet the SCL Protocol focuses solely on the main contractor's programme and does not, for example, mention any system for programming the activities of consultants.
- The SCL Protocol recommends agreement of a programme 'As early as possible in the project' but does not specifically recognise a pre-construction phase, instead envisaging a single programme describing how the main contractor 'plans to carry out the works'[367].

---

[364] Project case study 3, Appendix A.
[365] SCL Protocol (2002).
[366] SCL Protocol (2002), Section 1, 5.
[367] SCL Protocol (2002), Guidance Section 2.2, 35.

- The SCL Protocol hardly mentions the programming of 'design', alluding only briefly to it being part of 'all relevant activities' to be covered by a programme together with 'manufacturing, procurement and on-site construction'[368].
- The SCL Protocol Model Specification Clause envisages preparation of an initial programme 'Within two weeks of the award of the Contract' showing 'the first three months' work' and then a further full programme 'within four weeks of award'[369], although the Guidance Section states that ideally a draft programme should be submitted before works on site are commenced[370].

While advocating avoidance of disputes through early pre-planning, the timing of programme preparation recommended by the SCL Protocol appears to permit delays and gaps in finalisation of a programme sufficient to generate their own uncertainties and disputes. In addition, having permitted the late provision of a programme by the main contractor rather than recommending its agreement by all parties ahead of contract award, the SCL Protocol goes on to propose the use of liquidated damages or the reduction of interim payments as a remedy for late supply of the programme itself[371]. These are adversarial remedies which do not ensure the early creation and agreement of a programme, but only allocate blame for failure or delay.

**5.4.7  Programmes as additional contract documents**

In the structure of the preconstruction phase agreement as an effective process contract, the programme will be the spine of such a contract. Without it, there is the risk of early contractual commitments binding the parties on an open-ended basis and allowing delay in commencement of construction[372]. There is, however, the practical challenge of creating a contractual programme in the appropriate format and language and with the appropriate level of detail. The format can be a bar chart or a list, or a combination of both, and is likely to combine an architect's or project manager's design release schedule with a main contractor's information required schedule and procurement programme[373]. Whether set out as an annotated bar chart or a matrix or a combination of both, the preconstruction phase programme needs

---

[368] SCL Protocol (2002), Guidance Section 2.2.1.1, 36.
[369] SCL Protocol (2002), Appendix B, paragraph 2.1 and 2.2.
[370] SCL Protocol (2002), Guidance Section 2.2.1.3, 36.
[371] SCL Protocol (2002), Appendix B, paragraph 3.
[372] See also Chapter 9, Section 9.3.4 (Concerns as to conditionality).
[373] See also Chapter 8, Section 8.2.4 (The use of programmes by project managers).

to state each activity that is time-critical, the relevant deadline, the party or parties responsible and the preconditions on which it is dependent.

> Bewick Court, Newcastle upon Tyne (housing refurbishment): the preconstruction phase programme formed part of the contract and was set out as a list stating owners of each activity and dates for completion of each activity. For example, it stated the responsibilities of the project manager, the architect and the main contractor for identifying and recommending a cladding specialist subcontractor. This served as a clear reminder that all parties should assist in seeking a solution when that specialist later went into administrative receivership[374].

> A30 Bodmin/Indian Queens (newbuild road): the preconstruction phase programme formed part of the contract and was set out as a list identifying dates for completion of each activity. Ownership of each task rested with the main contractor unless stated otherwise, for example 34 activities requiring 'comments' or other inputs of the Highways Agency as client. Other activities denoted third party input outside the control of the team, which therefore created preconditions to the progress of subsequent team activities, for example an Environment Agency 'comment period' following a request for approval, and the 'Secretary of State's Decision' following the public inquiry[375].

### 5.4.8 Remedies for non-compliance with preconstruction phase programmes

Where a preconstruction phase programme is incorporated in a contractually binding agreement, what action should the parties be entitled to take in the event that any of the deadlines set out in a preconstruction phase programme are not adhered to? The only detailed provisions appear in PPC2000, which provides for use of a preconstruction phase programme called the 'Partnering Timetable' and includes a right of termination by the client in the event that the main contractor 'does not commence and continue to fulfil its responsibilities under the Partnering Contract in accordance with the Partnering Timetable'[376]. As regards the consultants who are also party

---

[374] Project case study 1, Appendix A.
[375] Project case study 7, Appendix A.
[376] PPC2000, Partnering Terms, clause 26.4(i).

to PPC2000, there is a right of termination by the client in the event of material breach, which could include their failure to adhere to the preconstruction phase programme[377].

There is also a further general remedy in PPC2000 whereby failure of the main contractor to adhere to an instruction of the 'Client Representative' (the project manager under PPC2000) which is in accordance with contract (and therefore could include an instruction to adhere to the preconstruction phase programme) allows the client (after following an appropriate procedure) to pay another party to carry out the relevant instruction and to recover any consequent additional cost from the main contractor[378].

Such remedies emphasise the significance of a programme that creates binding preconstruction phase commitments. It is therefore important that a preconstruction phase programme states only those commitments that the team members can honour, and for this purpose it should identify clearly:

- Agreed deadlines for preconstruction phase activities, identifying which team members are to undertake each activity;
- The correct sequence of preconstruction phase activities;
- The dependence of team members on previous activities by other team members;
- Any obstacles or third party influences that could delay or frustrate particular preconstruction phase activities.

The purpose of this chapter has been to demonstrate how early project processes can benefit from a clearer understanding as to the involvement of the client, as to the ways that team members communicate with each other and as to the deadlines to which they are all working. These are means to facilitate progress and avoid confusion at a time when the construction phase appointment of team members still remains conditional on the early project processes being satisfactorily completed.

---

[377] PPC2000, Partnering Terms, clause 26.3.
[378] PPC2000, Partnering Terms, clauses 5.3 to 5.5.

# CHAPTER SIX
# CONTRACTUAL AND NON-CONTRACTUAL PRECONSTRUCTION OPTIONS

## 6.1 Introduction

It is possible that the preconstruction phase processes described in early chapters may be implemented by the parties without any formal contractual or other obligations. However, it is also arguable that without a contract to record a clear understanding between the parties, the different members of a project team may not construe their respective responsibilities and deadlines in the same way.

There are a number of ways in which project teams can clarify their preconstruction phase commitments. These include:

- Creation of binding contractual arrangements through bespoke agreements such as framework agreements or through the use of standard form building contracts that provide for, or can be adapted to create, preconstruction phase agreements;
- Corporate integration of the project team members, for example through a joint venture;
- The use of less formal agreements such as letters of intent;
- Reliance on non-binding or even unwritten understandings.

## 6.2 Building contract options

### 6.2.1 Bespoke agreements and standard forms

It is possible for team members to create their own bespoke preconstruction phase agreement. However, the time and cost involved are unlikely to be justifiable in preference to use of an appropriate published standard form, except in circumstances where:

- The bespoke form is a prototype for a new published standard form of contract; or
- The bespoke form will be used extensively across a significant number of projects procured by the same client, for example under a large-scale framework agreement.

*Early Contractor Involvement in Building Procurement*

As regards new published forms, PPC2000 was trialled as a bespoke prototype on a number of projects, including the London Borough of Hackney's Nightingale Estate project[379]. NEC2 was first published as a 'Consultative Version' in order to encourage feedback from users.

As to framework agreements, these are considered further in Chapter 7 and Appendix C. They can include provisions that describe preconstruction phase processes for each project that falls within their scope and duration. Partnerships for Schools developed a suite of bespoke documents, including a Strategic Partnering Agreement, to govern its Building Schools for the Future Programme[380].

As regards the preconstruction phase processes set out in standard form building contracts, Appendix B comprises a review of the treatment of design development, two-stage pricing, risk management, communications, programming and team integration under various standard forms, namely GC/Works/1 Two Stage Design and Build, NEC3, PPC2000, Perform 21, JCT 2005 and JCT CE.

### 6.2.2 GC/Works

GC/Works/1 Two Stage Design and Build was published in 1999 and was the first published standard form to describe preconstruction phase processes. It was specifically created for two-stage tendering and envisaged the conditional appointment of a main contractor for the purpose of participation in design, converting to an unconditional appointment when such design, together with supply chain arrangements and prices, had been sufficiently developed[381]. Unfortunately, there is little evidence of the GC/Works/1 Two Stage Design and Build being utilised in practice[382].

### 6.2.3 NEC3

NEC3 comprises a complementary set of consultant appointments, main contracts and subcontracts, but does not include a preconstruction phase agreement or expressly provide for preconstruction phase processes. The NEC3 Professional Services Contract can be adapted to create a preconstruction phase agreement and includes risk management and communication provisions. However, as drafted, the NEC3

---

[379] See Project case study 6, Appendix A.
[380] See also Chapter 7, Section 7.5 (Frameworks and the Private Finance Initiative).
[381] See also Appendix B.
[382] RICS (2001) recorded no use at all of the GC/Works/1 Two Stage Design and Build form and RICS (2004) recorded one use of it, RICS (2001), 17, and RICS (2004), 18.

*Contractual and Non-contractual Preconstruction Options*

Professional Services Contract is a generic stand-alone consultant appointment not designed to be a conditional contractor appointment and not linked contractually to a subsequent construction phase NEC3 building contract. This may have adverse consequences in terms of its commercial attractiveness to a contractor asked to commit its expertise to preconstruction phase activities, and may also create complications for a team that is seeking to establish an integrated two-stage procurement process[383].

Alternatively, NEC3 can be preceded by a bespoke preconstruction phase agreement, whether for a single project or as part of a framework. This approach was adopted on the Eden Project Phase 4[384].

NEC recommend the use of their NEC3 Option E Cost Reimbursement Contract as a basis for establishing a preconstruction phase agreement[385]. However, this contract is not a conditional agreement linked in any way to a construction phase NEC3 contract and would need significant amendment to achieve appropriate links. NEC3 Option E also contains a number of clauses that are inappropriate in respect of preconstruction phase activities[386].

### 6.2.4 PPC2000

PPC2000 comprises a single multi-party contract with a complementary subcontract[387]. It is designed to be signed early in the preconstruction phase by the client, consultant, main contractor and certain subcontractors and suppliers in order to govern the following processes:

- Joint design development[388];
- Joint selection of remaining members of the main contractor's supply chain[389];
- Build-up of prices[390];

---

[383] See also Chapter 2, Section 2.1 (The conditional preconstruction phase agreement) as to conditionality and Chapter 9, Section 9.3.1 (Constitutional or regulatory constraints) as to public procurement issues.
[384] See Project case study 11, Appendix A.
[385] NEC Procurement and Contract Strategies, 20, 21.
[386] NEC3 Option E, see for example assumptions in respect of the risk register (clause 16.3), design (clause 21) and subcontracting (clause 26) as well as provisions in respect of taking over (clause 25) and testing (clause 40).
[387] See also Appendix B.
[388] PPC2000 Partnering Terms, clauses 8.3 to 8.12.
[389] PPC2000 Partnering Terms, clauses 10.3 to 10.9.
[390] PPC2000 Partnering Terms, clauses 12.5 to 12.7.

- Joint risk management[391];
- Agreement of a construction phase programme[392].

PPC2000 provides for these activities to be governed by a preconstruction phase programme known as the 'Partnering Timetable'[393]. It also provides that commitment to the construction phase of the project remains conditional upon the parties satisfying a series of agreed preconditions before then signing up to a 'Commencement Agreement'[394] confirming the readiness of the project to proceed on site.

### 6.2.5 Perform 21

Perform 21 comprises a complementary set of consultant appointments, main contracts and subcontracts[395], and provides for a preconstruction phase agreement separate from the construction phase contracts[396]. This agreement is intended to govern preparatory activities similar to those described in PPC2000, albeit that in Perform 21 they are described in less detail.

Perform 21, like PPC2000, recognises the principle that it may be beneficial to the project for main contractors, subcontractors and suppliers to be formally appointed alongside consultants during the preconstruction phase.

### 6.2.6 JCT 2005

JCT 2005 comprises a complementary set of main contracts and subcontracts, but without any corresponding consultant appointments until the publication of JCT CA at the end of 2008[397]. JCT 2005 does not provide for conditional arrangements during the period until the project is ready to proceed on site, with the limited exception of the design submission procedure contemplated by the JCT 2005 Major Project Construction Contract[398].

Separate preconstruction phase commitments independent from construction phase building contracts appear in the JCT 2005 Framework

---

[391] PPC2000 Partnering Terms, clauses 12.9 and 18.1.
[392] PPC2000 Partnering Terms, clause 6.2.
[393] PPC2000 Partnering Terms, clause 6.1.
[394] PPC2000 Partnering Terms, clause 15.1.
[395] See also Appendix B.
[396] Perform 21 PSPC 10 Prestart Agreement.
[397] JCT CA is the first JCT consultant appointment, stated to be intended for public sector clients.
[398] JCT 2005 MPCC, clause 12.

Agreement where it is contemplated that collaborative risk analysis will be undertaken prior to entering into specific 'Underlying Contracts' (i.e. building contracts)[399].

In late 2008 the JCT launched their Pre-Construction Services Agreement[400] for use between a client and a main contractor and their Pre-Construction Services Agreement (Specialist) [401] for use between a client or main contractor and a specialist contractor or subcontractor. These agreements are significant additions to the JCT suite of contracts as they are designed to provide for the preconstruction phase contributions of main contractors and specialists.

### 6.2.7 JCT CE

JCT CE takes the form of a series of substantially identical purchase orders which can describe the client/consultant, client/main contractor or main contractor/subcontractor relationships[402] and which can together be supplemented by a multi-party agreement dealing solely with risk/reward mechanisms[403]. It does not include an integrated conditional preconstruction phase appointment or expressly provide for preconstruction phase processes. A JCT CE contract can be adapted to create a preconstruction phase agreement, but would need amendment to describe preconstruction phase processes and to establish the preconditions governing award of a construction phase JCT CE contract. It is interesting that although JCT CE is not structured so as to create a conditional appointment governing preconstruction processes, its guide recommends early involvement of contractors and key specialists under 'a two-stage appointment with separate forms of agreement for a preconstruction stage for services (Stage 1) and for the construction stage (Stage 2)'. The JCT CE Guide also states that 'Normally a client will retain the option not to proceed to the construction stage so as to provide some commercial pressure on the contractor not to pitch his assessment of the target cost for the construction period too high'[404].

---

[399] JCT 2005 Framework Agreement, clause 15.1. The JCT Framework Agreement is considered further in Chapter 7 (Preconstruction commitments under framework agreements) and Appendix C.
[400] JCT PCSA.
[401] JCT PCSA(S).
[402] See also Appendix B.
[403] JCT CE Project Team Agreement.
[404] JCT CE Guide, Sections 37 and 38, 7.

## 6.2.8 Joint ventures

An alternative contractual structure which can secure main contractor commitment to preconstruction phase activities is the client's participation in the main contractor organisation itself through a direct shareholding in a joint venture. While this involves risk in respect of the client's investment and the commitment of additional client resources to participate in management, those clients with a significant flow of work can use participation in a joint venture to ensure shared information and joint commitment at all stages of each project. However, to put this commitment into practice will also require a further contractual relationship to be established between the joint venture company and its client or clients governing the delivery of the required services and works. This gives rise to the question of whether this further contract should document the implementation of project-specific preconstruction phase processes. A model of this type was used by Sheffield City Council who invested in a Limited Liability Partnership with Kier Support Services to undertake works of repair and refurbishment of all the Council's housing, schools and other public buildings, creating 'A partnership that will share the Council's core values, in particular its commitment to equal opportunities, social justice and regeneration'[405].

## 6.3 *Letters of intent*

In the absence of a formal preconstruction phase agreement, some clients will seek to secure preconstruction phase commitments through the use of a letter of intent. Letters of intent remain popular in the construction industry as a contractual half-way house by which the client can make a limited preconstruction phase commitment to its proposed main contractor.

Letters of intent are bespoke documents that are not provided for in any standard form building contract, hence their structure and content vary widely. Depending on their terms, letters of intent can have any of the following contractual effects:

- A letter of comfort with no contractual status;
- A preconstruction phase agreement governing limited conditional activities pending unconditional award of the construction phase building contract;
- An informal, but nevertheless unconditional, award of the construction phase building contract.

---

[405] Kier (2005), 4.

Generally, the purpose of a letter of intent is the second of the above. However, depending on its wording, difficulties may arise as to its enforceability and in ensuring that its implementation is governed by the relevant provisions of the building contract without the need to write an entirely new set of contractual provisions governing the letter of intent. It is possible that the terms of a letter of intent may have a 'negative contractual intention', but also on the contrary that the courts may 'hold the parties bound by the document', particularly where the parties have relied on it for a long time as a basis for their actions and payments[406].

Letters of intent are usually brief documents that do not describe preconstruction phase processes in any detail, being created primarily to meet a main contractor's need to secure a financial commitment from its client so as to underwrite urgent expenditure, for example on-site preparation or a long lead-in order for goods or equipment. Hence, they may not state exactly what activities are being undertaken or the date when they need to be completed.

In order to be enforceable but conditional, letters of intent need to adopt the characteristics of a preconstruction phase agreement, including provisions to deal clearly with each of the following:

- Timing and procedure for finalising the construction phase building contract, particularly if the position starts to change significantly from the documents referred to in the letter of intent[407];
- Procedures to agree on key issues left outstanding as at the date of the letter of intent, for example the date of start on site and date for completion, prices of particular elements of the works and finalisation of detailed designs[408];
- Relationships with consultants, for example interfaces with design consultants for the purpose of capturing the value engineering and buildability inputs of the main contractor[409];
- Timing of preconstruction phase activities[410];

---

[406] Chitty (2008), 2–125, 212.
[407] See also Chapter 2, Section 2.5 (Choices and contractual conditionality) regarding the need for a system to move through a series of choices until establishment of a final binding agreement.
[408] See also Chapter 2, Section 2.4 (The planning function of contracts) regarding planning functions whereby the conflicting interests of a negotiation process can be dealt with by establishing enterprise planning through, for example, subcontract tendering or by formalising the submission of business cases put forward by a team member.
[409] See also Chapter 2, Section 2.3 (Effect of the number of parties) regarding the challenges of reconciling the interests of multiple parties.
[410] See also Chapter 5, Section 5.4 (The role of binding programmes) regarding the need for agreed deadlines.

*Early Contractor Involvement in Building Procurement*

- Effect of preconstruction phase activities on any quoted fixed price, for example as to what activities may constitute variations[411];
- Limits on the main contractor's financial authority and arrangements for payment[412].

The popularity of letters of intent appears to reflect a deeply held belief in the industry that, whatever their contents, they are not full contracts and therefore they involve less risk for their signatories. Notwithstanding the reality of the risks described above, and the correspondingly greater clarity and lower risks of a full preconstruction phase agreement, the challenge of encouraging the industry to move away from the familiarity of the letter of intent towards a more thorough preconstruction phase agreement is considerable.

The vulnerability of letters of intent that are not tied to clear contractual terms was picked up in the Arup report to OGC when reviewing the Pre-Possession Agreement that governs early activities on site under PPC2000. Arup observed that:

> 'The Pre-Possession Agreement is a well thought out method of allowing works to be carried out whilst the documentation for the project is being developed. Providing a document to commence the project which is coordinated with the main contract processes and that prompts the parties to continue with developing the main contract documentation is superior to a stand-alone letter of intent'[413].

## 6.4 Non-binding arrangements

### 6.4.1 Non-binding protocols

Non-binding project protocols are bespoke documents intended to describe the processes and timetables governing project team members' contributions to the preconstruction phase of a project but without creating a contractual commitment to such processes and timetables. Such protocols are a means to accommodate the parties' reluctance to make early contractual commitments, but a fundamental problem remains: what is the benefit of agreeing something on a 'non-binding'

---

[411] See also Chapter 2, Section 2.7 (Risk and fear of opportunism) regarding the risk of opportunism where errors arise through limited foresight, and a lack of contractual clarity leads the parties to opportunistic behaviour so as to avoid the consequences of such errors.
[412] See also Chapter 1, Section 1.4 (Early contractor appointments and payment).
[413] Arup 2008, 47. The PPC2000 Pre-Possession Agreement is now renamed as the 'Pre-Construction Agreement', PPC2000, Appendix 3, Part 1.

basis if that means the parties cannot rely on each other when committing resources? In practice, supposedly non-binding project protocols create considerable uncertainty as they may inadvertently render the remainder of a contract unenforceable or may be enforced as part of such a contract or may simply be ignored alongside binding contractual commitments.

In *Birse Construction* v. *St David*, Judge Humphrey Lloyd made it clear that a partnering charter could affect the parties' substantive contractual rights[414]. On the other hand, Chitty noted that the courts will ignore 'verbiage, not intended to add anything to an otherwise complete agreement' and that if 'a meaningless clause' governs some vital aspect of the parties' relationship, then their vagueness may vitiate the entire agreement[415].

The use of non-binding protocols has been linked to the suggestion that in a partnering relationship formal contracts are no longer important and the parties can base such a relationship on a non-binding document known as a 'partnering charter'. This is usually a very brief document intended to capture headline statements of the agreed values, goals and priorities of a partnering team. An example is annexed to JCT Practice Note 4 which sets out a specimen 'non-binding partnering charter for single project' which is a blank form except for the statement that 'The team agree to work together on [the project] to produce a completed project to meet agreed client needs and meet agreed quality standards within agreed budget/price and agreed programme'[416]. This is no more than a summary of what needs to be set out in any event in the relevant project building contract and consultant appointments.

A partnering charter was never intended to be a detailed working document or to describe specific roles, responsibilities and relationships. JCT Practice Note 4 itself recognised that there was a place for a full partnering contract, and the JCT version of such a contract later emerged in the shape of JCT CE. To quote the CIC Guide that formed the foundation for PPC2000: 'While it is recognised that partnering charters have served a valuable role, the time is right to see a fully integrated approach, so that the relationships and processes required for effective partnering are not at odds with the contractual roles and relationships of partnering team members'. The CIC Guide also stated 'For the avoidance of doubt what we are talking about is a legally binding contract and not a non-legally binding charter or any equivalent'[417].

---

[414] *Birse Construction Limited* v. *St David Limited* (2000) 1BLR57.
[415] Chitty (2008), 2-144, 226.
[416] JCT Practice Note 4, 9.
[417] CIC (2002), 12.

### 6.4.2 Unwritten understandings and partnering

Reliance on unwritten understandings, for example developed at meetings and in ad hoc communications, recorded only in minutes and correspondence, has an appeal when preconstruction phase processes are moving fast and when it might seem bureaucratic to stop and write things down in a formal document. However, this approach has the following risks:

- Misunderstandings – anyone who has read the minutes of a meeting and disagreed with them will know that reliance on meetings (even minuted ones) as the basis for action can lead to wasted time and disappointed expectations[418].
- Inefficiency – without a written agreement, there will be a need for team members to use additional meetings, correspondence and other means to remind each other what they are doing. This requires extra time and resources, which cost money[419].
- Bad faith – without a clear understanding as to agreed activities and commitments, one party could deny what was agreed and cheat another out of its expected entitlements. For example, in *Baird Textile Holdings* v. *Marks & Spencer* one party attempted to deny the existence of a contractual relationship based on the principle of 'partnership' rather than clear written terms, and it was reported that Judge Morison 'dismissed the claim insofar as it was based on contract but directed that it proceed to trial insofar as it was based on estoppel'[420].
- Excessive caution – lack of clarity can give rise to a lack of confidence as to what level of commitment has been established. This may reduce the parties' willingness fully to honour their commitments or to give them priority when compared to contractual obligations on other projects. This in turn may slow down the progress of preconstruction phase activities[421].

---

[418] See also Chapter 2, Section 2.10 (Alignment of different interests) regarding the ways that different parties read and react to signals differently, and the need for cooperative adaptation to clarify agreed collective interests.

[419] See also Chapter 2, Section 2.8 (Conditional relationships without full consideration) regarding the need for clear terms to encourage commitment in the absence of full reciprocity.

[420] *Baird Textile Holdings Limited* v. *Marks & Spencer Plc* (2001) EWCA Civ274. See also Chapter 2, Section 2.7 (Risk and fear of opportunism) regarding the risk of opportunism where the parties are dependent on negotiation rather than agreed processes.

[421] See also See also Chapter 2, Section 2.8 (Conditional relationships without full consideration) and Section 2.10 (Alignment of different interests) regarding some of the challenges in creating preconstruction phase agreements. See also Chapter 9, Section 9.3.4 (Concerns as to conditionality).

The possibility of working on a project without the support of a contract was considered by Egan, who concluded that building contracts frequently had a negative effect on the success of projects and suggested that the industry would be better off without them altogether. His report stated that 'If the relationship between a constructor and employer is soundly based and the parties recognise their mutual interdependence, then formal contract documents should gradually become obsolete'[422]. This Egan statement has been extensively quoted, but it was not explained or demonstrated in the Egan Report and was almost immediately called into question by the findings in *Baird Textiles Holdings v. Marks & Spencer*.

It has also been suggested that partnering can be undertaken with no contract at all, and that excessive attention to contractual matters can undermine the working culture required for successful partnering. Bennett & Pearce inferred that a formal contract of any kind is not always required or appropriate for partnering. They stated that 'Negotiating the terms of a formal contract tends to destroy partnering attitudes. Working to rules and procedures defined in a standard form of contract inhibits partnering behaviour'[423]. These observations may be intuitively attractive, and clearly it is preferable if day to day activities are guided by professionalism and personal values rather than by written rules and procedures. However, Bennett & Pearce did not recognise the commercial dangers of placing reliance on unwritten, unprogrammed, spontaneous performance of important design, procurement and construction activities that are subject to interlocking deadlines.

However, if significant aspects of a relationship between the parties are not covered in a written contract, then this suggests naivety as to the importance of clarity when entering into complex commitments. Cox & Townsend expressed doubts as to whether 'those who simplistically believe that collaboration based on trust alone, without an effective hierarchy of control in the relationship, can achieve improvements in construction outcomes'[424].

Arrighetti *et al.* challenged the view of project partnering that draws a distinction between, and seeks to separate, partnered processes and contractual relationships. They also questioned whether 'co-operation based on self-interest may emerge without the need for the intervention of the legal system', and suggested that instead 'the more complex types of contractual agreement may provide a foundation for "systems trust" by formalising shared expectations and assumptions of what constitutes accepted behaviour'[425].

---

[422] Egan (1998), 33.
[423] Bennett & Pearce (2006), 41.
[424] Cox & Townsend (1998), 333.
[425] Arrighetti *et al.* (1997), 175.

It is a source of serious concern that partnering has often been portrayed as a kind of parallel universe where the relationships and activities of the parties do not require the rigour or discipline that is present in other commercial relationships and activities. It is this view that often fuels the illusion that an unwritten understanding will suffice as a basis for partnering. Yet serious commercial and safety issues are at stake, and it is highly unlikely that an unwritten understanding will suffice as the basis for any relationships or activities governing any building project whether partnered or not. Without a written contract the parties cannot be sure that they have established clearly the interplay and systems of control governing quality, cost and time that are necessary for successful project completion.

A variant adopted in early partnering projects was a twin track contractual structure, whereby the parties adopted relational contracting as a basis to pursue collaborative working, but also entered into a conventional standard form building contract to fall back on if their relationship did not provide them with the desired results. Barlow *et al.* found in their research that even in well-developed partnering relationships there was still felt to be a need to rely on underlying contracts, and one of their suppliers commented 'No matter how many games of golf we've played or how many lunches they've taken you out for, I wouldn't be happy accepting an appointment without something in writing'[426]. As to the type of contracts used by partnering teams at the time of the research of Barlow *et al.* in 1997, they found that 'Standard contracts such as the JCT80 were nearly always used despite the admission by some interviewees that they were negative in their structure and displayed objectives that were the opposite to the principles of partnering'[427].

The required investment of time and money required in a construction project, and the need for clarity and certainty as to project team members' roles and responsibilities, suggest that reliance on an unwritten understanding needs to be considered very carefully before being adopted as a model for any aspect of procurement and project management. While it could be argued that effective communication systems and well-developed collaborative working allow the parties to rise above their contractual rights and obligations and to achieve a solution more beneficial to the project[428], a written agreement is still needed to establish a commercially sound starting point. The risk of confusion when non-binding arrangements are combined with other contractual

---

[426] Barlow *et al.* (1997), 34.
[427] Barlow *et al.* (1997), 34.
[428] See also Chapter 5, Section 5.3 (The role of communication systems) and Chapter 8, Section 8.3 (Preconstruction phase agreements and partnering).

arrangements was illustrated in the judgment of Judge Humphrey Lloyd in the case of *Birse Construction* v. *St David* where he stated 'In appropriate circumstances the provisions of a partnering agreement could be taken into account when interpreting the terms of an underlying contract'[429].

In practice and in the light of forms of contract now available, a written agreement can be considered an integral part of an effective communication system and a basis for collaborative working rather than offering only a fallback position.

## 6.5 Benefits of contractual clarity

Where preconstruction phase processes are recognised as having benefits, the absence of a preconstruction phase agreement can have the following consequences:

- Lack of clarity of mutual commitment as to what preconstruction phase activities are expected from the main contractor and what (if anything) it will be paid for those activities, thereby increasing the risk of misunderstandings and disputes;
- Delay in preconstruction activities while the nature, timing and value of main contractor input is established, thereby losing the benefit of early project planning;
- Confusion of preconstruction phase input by the main contractor with negotiation of a second stage unconditional construction phase building contract, thereby losing the enforceability of preconstruction phase commitments.

It is therefore arguable that formal preconstruction phase agreements, preferably using published standard forms, offer the most sound means of describing and programming preconstruction phase activities. Their role was hinted at by Banwell in his recommendation that:

'The breaking down of the present dividing line between design and construction, and the recognition of the fact that contractors are sometimes able successfully to take part in the preparation of a project, will mean that, in some cases, changes will have to be made in the time honoured procedures under which contracts are let'[430].

The 1998 CIRIA report recommended that the main contractor should be appointed 'under [a] suitable contract to contribute to project

---

[429] *Birse Construction Ltd v. St David Limited* (2000) 1 BLR 57.
[430] Banwell (1964), 6, Section 2.11.

development', including full involvement in methods, design and specification ahead of finalising contracts for the construction phase.[431] As to the form of such preconstruction phase agreement, the CIRIA report stated that the main contractor's role is 'more comparable to consultancy than the traditional contracting role and might require a separate, bespoke agreement'[432]. That report appeared prior to publication of any of the standard forms reviewed in this book.

This chapter has outlined a variety of options available when embarking on early project activities with a main contractor and its subcontractors and suppliers. It has noted the differing approaches taken in various published forms of building contract and has questioned the value of letters of intent and the effectiveness and appropriateness of arrangements that are not legally binding.

---

[431] CIRIA (1998), Figure 3.2, Model 1, 32.
[432] CIRIA (1998), 37.

# CHAPTER SEVEN
# PRECONSTRUCTION COMMITMENTS UNDER FRAMEWORK AGREEMENTS

## 7.1 *Commercial attraction of frameworks*

Properly organised and resourced preconstruction phase processes require commitment and investment by the parties. Although these preconstruction phase processes are for the benefit of the project in any event, it will be easier to attract the required commitment and investment if the project is part of an ongoing relationship between the client and the main contractor, consultants, subcontractors and suppliers. This chapter will look at the ways in which framework agreements governing more than one project can constitute conditional preconstruction phase agreements.

Framework agreements governing a series of projects can motivate the parties by identifying their common commercial goals and objectives, namely the continued course of business based on accepted levels of performance and pricing. However, as regards the preconstruction phase provisions of a framework agreement, these may need to remain more flexible than those agreed for a single project as the parties will be more likely to deal with much of the detailed planning of individual projects only when each project is identified and initiated. In these circumstances, where complete description of preconstruction activities is not possible because the relevant facts are not yet known, such a framework agreement is in part relational as it serves to give structure to the parties' relationship and record their common expectations. It also should establish clearly the mechanisms by which the parties will make the decisions necessary to implement individual projects and to agree the rewards for such projects[433].

The contractual focus may at first be less on agreeing a detailed plan of action during the preconstruction phase of a specific project and more on the establishment of processes and procedures that will

---

[433] See Milgrom & Roberts as to the need 'to structure a relationship and set common expectations, and ... establish mechanisms that will be used to make decisions and allocate costs and benefits', Milgrom & Roberts (1992), 132.

themselves be applied to formulate a detailed plan for each project, so as to save the time and cost that would otherwise be incurred in agreeing such processes and procedures each time a project is initiated. MacNeil observed that a strong relationship between the parties acts as a starting point for moving into the substantive planning required for particular projects[434], but that is not to say the framework governing the overall relationship cannot be set out in clear written terms.

As a framework arrangement gives rise to the prospect of a series of projects, the safeguarding of a party's reputation linked to the award of such future projects is likely to assist the voluntary operation of the required relationship. In these circumstances, it has been argued that the parties will be prevented from letting each other down or abusing their relationship by their concern that this will lead to them getting a bad reputation, and that the incentive for that opportunistic behaviour is offset because its short-term gains are outweighed by its long-term reputational damage. Milgrom & Roberts noted that 'the concern with getting a bad reputation that reduces future possibilities for profitable transactions can limit reneging', and thereby 'removes the incentives for opportunistic behaviour by creating a cost offsetting the short-term gains of opportunistic behaviour'[435].

Even the sceptical Duncan-Wallace recognised the commercial motivation for contractors to behave less opportunistically when additional projects are on offer. Duncan-Wallace commented in *Henry Boot Construction Ltd* v. *Alsthom Combined Cycles Ltd* [1999] Build, L.R. 123:

> 'Nor can there be any doubt that in a reasonable world, let alone one illuminated by the "partnering" or "good faith" or "cooperation" principles to which contractor influences as exemplified by the NEC contract and the Latham Report attach such importance (at least where advantage to their interest is contemplated), as also one where a more straightforward commercial desire of contractors to secure goodwill or further possible contracts in an owner's gift may be present, advantage would not have been taken of these major mistakes in calculating a last-minute quoted price so as to extrapolate obviously inappropriate and excessive prices'[436].

However, there is a risk that a loosely worded framework agreement may be unenforceable as a mere agreement to negotiate[437]. Hence, it is just as important under a framework agreement as under a single

---

[434] See MacNeil (1981), 1044.
[435] Milgrom & Roberts (1992), 139.
[436] Duncan-Wallace (2000), 2.
[437] See Chapter 2, Section 2.1 (The conditional preconstruction phase agreement) regarding negotiation and enforceability.

project contract that the preconstruction phase machinery leading up to the unconditional award of construction phase building contracts is clearly set out.

> Job Centre Plus (office refurbishment programme): two clients and 14 main contractors invested in agreeing a common set of preconstruction phase processes that were applied to each of a series of over 960 projects during a three-year period. Corresponding frameworks were created between the clients and key subcontractors and suppliers under which works and supplies were called off by individual main contractors on their respective projects. Operation of the processes described in these frameworks enabled the team to achieve savings of £244m against a projected cost of £981m[438].

## 7.2  The relationship between frameworks and partnering

Commentators have observed that the benefits of partnering are better achieved through long-term relationships where the same project team works on a series of successive projects, acquiring familiarity with each other's ways of working, increased trust in each other's integrity and practical lessons that are of benefit to later projects. A 1991 NEDC report stated that:

> 'On traditional single projects, personnel from different organisations in the project chains are unfamiliar with each other and there may only be a limited level of trust … The long-term nature of partnering means that parties on all sides are familiar with the project requirements and the level of trust which has been built up over a series of projects is there from day one on the next project'[439].

Whatever the level of such familiarity and trust, the commercial implications of joint working on successive projects are no different from those applicable to a single project, and a written agreement remains important for the same reasons.

The NEDC view in 1991 was that a long-term relationship needs to be based on commercial foundations, and observed that 'The ability to provide a significant core work programme and then to retain a core team is essential to the maintenance of any ongoing partnering arrangement'[440]. As to duration, their recommendation was that 'a period of

---

[438] Project case study 10, Appendix A.
[439] NEDC (1991), 34.
[440] NEDC (1991), 12.

five years should be taken into consideration as an absolute minimum'[441].
A long term commitment, whether for five years or any other significant period, is only justifiable if it continues to deliver better value than alternative single project arrangements. Whatever the contractual structure, contractors who have the benefit of a framework agreement can expect to be measured according to agreed performance targets and will need to adopt the machinery necessary to demonstrate that such targets have been met.

To establish strong commercial foundations and a duration on which the parties can rely when investing in new ways of working, it is suggested that a framework agreement is required that provides for clear preconstruction phase processes. These processes could, for example, be set out in conditional preconstruction phase agreements that form part of each project contract, using a model annexed to the framework agreement.

> Whitefriars, Coventry (housing refurbishment programme): the client and two main contractors set up a joint framework agreement to govern a five-year programme of housing work, which included a model two-stage project contract providing for preconstruction phase processes governing completion of the client brief, price framework and construction programme as preconditions to start on site. This enabled the three parties together with certain specialist subcontractors to work together on a partnering basis to agree and implement ways to save costs and increase efficiency, for example by means of a common supply chain and a shared training and employment initiative. The resultant efficiencies reduced a five-year programme to four years and saved the client 10% of its £240m expected cost[442].

## 7.3 Frameworks and preconstruction phase processes

A framework agreement can itself set out preconstruction phase processes so that individual building contracts for specific projects are created only once those preconstruction phase processes have been satisfactorily completed. In this way, the framework agreement becomes the preconstruction phase agreement under which joint design development, joint price and supply chain development, joint risk management and joint agreement of a construction phase programme are undertaken.

---

[441] NEDC (1991), 20.
[442] Project case study 9, Appendix A.

*Preconstruction Commitments Under Framework Agreements*

> Eden Project Phase 4 (newbuild leisure programme): the main contractor and a team of consultants were appointed under a multi-party framework agreement that contained provisions for joint preconstruction phase design development, pricing, risk management and programming prior to award of each construction phase contract. This attracted the team's early commitment to joint design of the innovative Education and Resource Centre (the 'Core') as well as in a number of smaller projects[443].

Whether the preconstruction phase processes are set out in the framework agreement itself or in the building contracts governing successive projects, framework agreements can establish the contractual commitment of the client, the main contractor and other parties to implementation of a programme of work. They can also set out the procedures by which that commitment will become unconditional in relation to successive projects.

## 7.4 Published forms of framework agreement

Until 2005 there were no published forms of framework agreement, and then two came along at once: the JCT 2005 Framework Agreement and the NEC3 Framework Contract. Neither document describes a system for awarding building contracts for individual projects or any preconstruction phase processes for such projects[444].

The JCT 2005 Framework Agreement was not in fact a framework agreement at all. It did not set out the scope or duration of the parties' relationship, but described instead collaborative values and activities that it encouraged the parties to apply to whatever projects they undertook together. The majority of these provisions were retained in the JCT 2007 Framework Agreement, and a number of the stated values and activities appear to clash directly with the corresponding provisions of the JCT 2005 standard form building contracts.

For example, clause 12.1 of the JCT 2007 Framework Agreement encourages a communications protocol, a potentially valuable document governing 'clear and effective communication and the dissemination and ready availability of information'. However, this protocol is stated to be separate from and without affecting the notice and communication requirements of the JCT 2005 building contracts pursuant to which each project is constructed (the 'Underlying Contracts'). Will

---

[443] Project case study 11, Appendix A.
[444] See Appendix C.

project participants make any serious effort to create or adopt a communications protocol if their contractual relationship on each project remains governed by a different set of communication procedures? Similarly, clauses 19 and 20 of the JCT 2007 Framework Agreement contain provisions dealing with 'Early Warning' and a 'Team approach to problem-solving', but again detach these collaborative processes from the more traditional dispute resolution provisions set out in the JCT 2005 Underlying Contracts. The clause 19 early warning provision expects a party to notify a matter 'likely to affect the out-turn cost or programme or the quality or performance of any Tasks', but specifically states that this is 'Without in any way detracting from or affecting the particular notice requirements of the Underlying Contracts'. Unless an early warning provision is connected directly to operation of the Underlying Contracts, it is hard to imagine how a party would be persuaded to give any early warning at all. For example, if the main contractor is aware of a problem in its own organisation or supply chain likely to affect an agreed date for completion of a project, why would it give early warning under the JCT 2007 Framework Agreement if this results in the client claiming liquidated damages in accordance with its entitlement under a JCT 2005 Underlying Contract[445]?

The JCT 2007 Framework Agreement introduced a system for call-off of particular 'Tasks' with provisions for these to be subject to a form of 'Enquiry' generating a response that would then lead to an appropriate 'Order' creating a specific project contract. However, the JCT 2007 Framework Agreement still omits any preconstruction phase processes that would involve a framework contractor (or any of its subcontractors/suppliers) in design development, in joint procurement of subcontractors/suppliers or in joint programming. Hence, the opportunity for methodical early contractor involvement in a project under the JCT 2007 Framework Agreement is neglected and the focus remains on the submission of prices and proposals that (subject to approval) lead immediately to a construction phase building contract.

One area where the JCT 2007 Framework Agreement does recognise early contractor involvement is in its reference to 'Risk assessment and risk allocation', where it is envisaged that the contractor (as part of the process of responding to an Enquiry) will work with other prospective project participants in 'collaborative risk analysis'[446]. However, it is interesting that while the JCT 2005 Framework Agreement contemplated in an equivalent clause the possibility of amended provisions in a project contract resulting from such joint risk analysis[447], this flexibility was removed two years later in the JCT 2007 Framework Agreement.

---

[445] See also further examples in Appendix C.
[446] JCT 2007 Framework Agreement, clause 14.
[447] JCT 2005 Framework Agreement, clause 15.

The inference is that whatever collaborative risk analysis is done by the prospective contractor in responding to an Enquiry, it is unlikely to affect the terms on which that contractor is expected to commit to the construction phase of the relevant project.

The NEC3 Framework Contract states the scope and duration of the relationship and also recognises the need for procedures whereby projects are awarded, but it provides no guidance or examples of how these procedures might be structured. It also requires creation of a separate NEC3 Professional Services Contract to govern any instruction to provide advice on a proposed work package on a time charge basis[448]. The NEC3 Framework Contract requires substantial additional drafting to fulfil the purposes of a framework in governing the award of a series of projects and in clarifying and controlling the preconstruction phase processes necessary before any project can be implemented on site.

## 7.5 Frameworks and the Private Finance Initiative

The considerable time and cost involved in concluding Private Finance Initiative (PFI) transactions is a powerful incentive for the Government and for PFI providers to group such transactions under framework agreements. Equally important is the concern that lessons learned during the intensive preparatory stages of earlier PFI transactions can be lost on later PFI transactions unless captured in development of preconstruction phase processes for later transactions implemented pursuant to a framework.

Partnerships for Health sought through its LIFT (Local Improvement Finance Trust) initiative to establish a framework whereby the initial transaction formed the basis for successive further transactions involving the same provider and an increasing group of health sector clients in a specific geographic location, corresponding to their developing needs over the life of the relationship. For this purpose it created a type of framework agreement (the 'Lease Plus Agreement') spanning the leasing and procurement of works in successive premises[449]. Similarly, the Building Schools for the Future (BSF) programme created a framework (the 'Strategic Partnering Agreement') to govern the procurement of a series of school projects, comprising a number of PFI transactions and a number of design and build projects. The Building Schools for the Future initiative specifically provides for successive PFI projects and individual design and build projects to be undertaken by an SPV, the 'local education partnership'. Shareholders of the local education

---

[448] NEC3 Framework Contract, clause 11.2(5).
[449] LIFT (2006).

partnership include the client and the provider, the latter comprising a vehicle in which the building contractor (as in the case of a PFI model) is likely to have an equity investment.

Although the preconstruction phase preparation for the first BSF project or projects is the subject of a public procurement procedure, the client and the provider enter into a Strategic Partnering Agreement governing the following joint preconstruction phase processes preparatory to all further projects:

- Establishment of feasibility, including surveys and investigations;
- Development of new project proposals including building solutions and design proposals through various stages of detail, incorporating contributions from a full range of design consultants;
- Indicative costing leading through to a detailed financial model, prices to be prepared on an open-book basis with proposals for development and management of supply chain members;
- Obtaining third party consents;
- Review of relevant risks;
- Preparation of a proposed programme[450].

The BSF Strategic Partnering Agreement contains a full range of preconstruction phase processes considered in this book, allowing the parties to apply them as preconditions to the award of successive design and build projects and PFI projects. Partnerships for Schools remain involved throughout a BSF programme to sign off stages of new project approval, in order to ensure at each stage that the agreed preconstruction phase processes have been carried out in accordance with their published guidance on the approved procedure for new projects, and in order to ensure that such agreed processes (set out in a *New Project Protocol*) are reviewed and updated as the partnership matures.

## 7.6  *The impact of frameworks on changing behaviour*

Efficient working among the members of a multi-party team is dependent on reasonable behaviour sufficient to generate trust so that the parties are willing to allow sensitive information to be shared among project team members. Colledge linked this sharing of information with relational contracting and observes that, while shared information achieves commercial value for a single project and the parties involved in terms of time, cost and quality objectives, the value of effective teamworking and the long-term benefit that creates for future projects will be more significant. She suggested that the commercial relation-

---

[450] PfS (2008), Schedule 3 New Project Approval Procedure.

ships formed through relational contracting 'not only foster mutual trust, but also facilitate the sharing of knowledge and information to generate innovation and value for the parties to the relationship'[451].

The benefits for future projects can only be fully realised if the same project team members are able to work together on such future projects under a stable framework, and the existence of a framework agreement is a significant factor in motivating the type of behaviour that leads to the required shared information. It clarifies the systems by which successive projects will be planned, designed and built, it describes the investments and rewards expected by the parties and it sets out the agreed measures of their performance[452].

The prospect of successive projects is a major incentive for collaborative working, but a system of performance measurement justifying the continued award of such projects is a commercial necessity for the client. Hence, it is surprising that the JCT 2007 Framework Agreement refers to performance to be measured against agreed key performance indicators, but does not link that measured performance to the award of future projects[453].

A framework agreement alone is not sufficient to change behaviour. Training of project teams in new ways of working and ensuring that all parties understand the links between the operation of the framework and the implementation of individual projects, are also essential preparatory processes[454].

> Project Y (newbuild and refurbishment schools programme): having established its framework, the client did not allocate sufficient time or resources to training its project teams:
>
> - in the way the framework should be used so as to incentivise the contractor to improve its preconstruction phase design and cost proposals; or
> - in consistent application of the joint new preconstruction phase processes required for each project.
>
> This lack of training led to confusion, inconsistent approaches between project teams, lack of joint preconstruction phase activities and a slow start to the programme of work[455].

---

[451] Colledge (2005), 32.
[452] See also Chapter 8, Section 8.3.3 (Features of partnering) regarding the views of Bresnen & Marshall as to the need for a commercial driver to achieve changes in the way that project team members work together.
[453] JCT 2007 Framework Agreement, clause 21.
[454] See also Chapter 9, Section 9.5 (Education and training).
[455] Project case study 12, Appendix A.

Campbell & Harris noted that the classical contracting model assumes that the contracting parties would rapidly change their positions if a different set of circumstances would give them a chance to increase their profits over and above the profits available by performing the existing contract[456]. They envisaged that in single project relationships either contracting party will exploit changed circumstances to the disadvantage of the other contracting party, and they suggested that a framework relationship is required to overcome this inclination. The reasoning is that, in the long-term contracts created by frameworks, exploitative behaviour is less likely because the parties will calculate that shifting towards such behaviour offers less reward than preserving a long-term contractual relationship. Taking a long-term view can secure honourable and constructive behaviour even where short-term dishonourable behaviour could give rise to an immediate profit, provided that the long-term view remains of greater commercial benefit.

In examining long-term contractual behaviour, Campbell & Harris suggested that efficient long-term contractual behaviour 'must be understood as consciously cooperative' and that, despite a range of collaborative or individualistic attitudes, 'those parties which contract efficiently act cooperatively' in a manner analogous to a partnership[457]. However, they considered that this cooperation is 'manifested in trust and not in reliance on obligations specified in advance', specifically because in a long-term cooperation it is not possible to specify all expectations in advance, and that as a result the parties 'accept a general and productively vague norm of fairness in the conduct of their relationship'[458]. I accept the need to develop and demonstrate evidence of such fairness, but I would argue that there are also rules and procedures that should be set out in a framework agreement to create a clear and bankable understanding between the parties that should also assist the development of trust and fairness in their dealings.

As a matter of commercial logic, the continued prospect of new work is the strongest motivation for the parties to overcome doubts regarding the need for change in methods of working and to avoid reverting to a short-term, adversarial stance when faced with an obstacle or difference of opinion[459]. It is not difficult to see why contracting parties would invest more by way of joint working to achieve improved results, including the sharing of information that they might otherwise want to retain, if there is the reasonable expectation of greater long-

---

[456] Campbell & Harris (2005), 5.
[457] Campbell & Harris (2005), 6.
[458] Campbell & Harris (2005), 6.
[459] See Chapter 9, Section 9.4.3 (Industry conservatism) regarding reluctance to change.

term benefits. Campbell & Harris perceived that self-interested behaviour can lead to conscious adoption of a cooperative approach where this is perceived to be the best long-term strategy, taking into account the following components:

- Analysis of the benefit of the relationship and assessment of the value that other parties place on the relationship;
- The incentive to continue the relationship combined with the disincentive against terminating the relationship;
- The expectation of an undefined share in the joint benefits that can be generated by the relationship (provided that such share is expected to be greater than any gain achievable by the party proceeding alone);
- The risk of losing specific investments made in the relationship and the potential cost of developing and investing in a new alternative relationship[460].

However, the establishment of these benefit analyses, incentives and commercial expectations in a long-term relationship requires written terms that the parties can interpret reliably. By the same logic as applies to a single project, if the parties' rewards are to be proportionate to their efforts, then there is a commercial rationale for new behaviour and additional commitments to be spelled out in writing.

A framework relationship will thrive so long as each party's analysis concludes that its continuation is more advantageous than its termination. Campbell & Harris described the thinking of such parties in a framework as 'I calculate that I shall be better off in the longer term if I continue my relationship with you instead of terminating it; and I also estimate that you similarly have calculated that you will be better off if the relationship continues'[461].

Those who expect to use framework agreements to bind other parties to an exclusive relationship, irrespective of whether original expectations are realised, may be disappointed by Campbell & Harris' analysis. Although a framework agreement as a process document can do much to ensure the efficient completion of joint preconstruction phase activities on successive projects, it can only continue to enhance commitment and cooperation by the parties if it contains machinery for them regularly to reaffirm their original underlying commercial calculations and estimations, namely whether their expectations of a steady flow of work (from the client) and measurable improved performance (by the contractor, consultant, subcontractor or supplier to whom the framework is awarded) continue to be fulfilled.

---

[460] Campbell, & Harris (2005), 23.
[461] Campbell & Harris (2005), 23, 24.

This chapter has demonstrated that an important contractual means of capturing early project processes is through the creation and implementation of framework agreements, where there is the potential for the same team to work together on more than one project, and that such agreements can also embed and measure the continued commercial motivation for improved performance of preconstruction phase activities.

# CHAPTER EIGHT
# PROJECT MANAGEMENT AND PROJECT PARTNERING

## 8.1 Introduction

Whether or not early project processes are governed by binding contracts and whether these govern a single project or a framework relationship, the efficient management of these processes will be fundamental to their success. This chapter will examine the role of project management in the early stages of a project and will focus in particular on the collaborative approach to project management known as partnering.

## 8.2 Preconstruction phase agreements and project management

### 8.2.1 What is project management and when does it start?

Project management is defined in BS 6079 as 'The planning, monitoring and control of all aspects of a project and the motivation of all those involved to achieve the project objectives on time and to cost, quality and performance'[462]. The role of 'project manager' is defined by Burke as having 'the single point responsibility to co-ordinate multi-disciplinary projects'[463]. Burke saw the project manager as responsible for developing the plan by which the project can be tracked and controlled to ensure that it meets its agreed objectives[464]. It is important, therefore, to consider the perspective of the individual or organisation that will have this responsibility as it will directly influence the treatment of preconstruction phase processes and the timing and nature of contractor involvement in those processes.

The purpose of project management was described by Lock as a system for foreseeing or predicting as many risks and problems as

---

[462] British Standard 6079 defining Project Management.
[463] Burke (2002), 275.
[464] See Burke (2002), 8.

possible in relation to a project and then planning, organising and controlling the activities required to overcome such risks and problems so that the project is completed successfully[465]. These are demanding responsibilities that span both the preconstruction and construction phases of a project. However, Burke noted that only in the 1980s did the emphasis of project management shift to focus more on the preconstruction phase of the project. He recognised that it is during the preconstruction phase of the project that the needs of stakeholders can be analysed, that project feasibility can be assessed, that value management can be encouraged to ensure that the project is being approached in the correct way, and that project risks can be assessed[466].

It is therefore during the earlier phases of a project that there is the greatest potential for good project management to bring added value, as this is the time when there is greatest opportunity for the efficient management of changes. Later introduction of changes, including those resulting from design error, will become increasingly expensive as the project progresses, and Burke stated that 'The ability to influence the project, to reduce project costs, build in additional value, improve performance and increase flexibility is highest at the very early conceptual and design stages'[467]. Burke used these arguments to support the early appointment of a project manager. The same arguments can be used to support the early appointment of main contractors and specialist subcontractors and suppliers, and such early appointments will require project management to programme and coordinate and to monitor progress of the agreed activities.

Despite increasing emphasis on the importance of the client's role in project procurement[468], active involvement by the client is not a substitute for effective project management, although the two may become intertwined in circumstances where the appointed project manager is an officer of the client rather than an external consultant. Local authorities, for example, frequently name one of their officers in a building contract to fulfil project management functions.

However, in a two-party building contract such as JCT 2005 or NEC3, the naming of an officer as architect/contract administrator or project manager does not clarify the nature and scope of their agreed activities, as there is no separate form of consultant appointment made between the client and its own officer that states these activities in detail and can thus clarify them to the lead designer, main contractor and other project team members. A particular cause of concern where the client appoints its own officer as project manager will be a lack of objectivity

---

[465] See Lock (2000), 3.
[466] See Burke (2002), 30.
[467] Burke (2002), 31.
[468] As described in Chapter 5, Section 5.2 (The role of the client).

whenever that party is called upon to exercise its judgement, for example regarding a valuation or the progress or adequacy of particular work[469]. Clarification of the services provided by an in-house project manager, and its exact relationship with the client, may therefore provide comfort to the main contractor and other team members and assist the efficient administration of the contract.

In this respect, the multi-party structure of PPC2000 provides an advantage as it is possible for the client organisation to enter into the contract in two capacities, that of 'Client' and 'Client Representative' (i.e. the project manager), and to set out in detail the project management services offered to the project team as a whole through the functions delegated to the officer who acts as Client Representative.

## 8.2.2 The role of the project manager in establishing a procurement strategy

The first responsibility of the project manager will be to decide or advise on the procurement strategy for the project. This is the point at which a decision will be taken by the project manager, or by the client on its advice, as to whether the prospective main contractor and any subcontractors and suppliers will be appointed early enough to participate in preconstruction phase processes. Burke cited the project manager 'Not working closely with the client' as a common reason for project failure[470]. It is important therefore that the project manager's own appointment requires it to advise appropriately, and relevant provisions in the published standard forms are not consistent.

The 1998 Association for Project Management (Standard Terms of Appointment (APM 1998)) described the project manager's responsibility for establishing a procurement strategy and for this purpose to:

- 'Advise on the most appropriate work procurement strategy for the Project;
- Advise on the procurement of Consultants, Contractors and others;
- Define responsibilities, lines of communication, reporting and authorization procedures between the parties for the Project and communicate these to every party;
- Advise the Client on obtaining appropriate specialist input, including terms of contracts, risk allocation, insurance, warranties and bonds'[471].

---

[469] See for example JCT 2005 SBC/Q, clause 2.3.3 which recognises that certain materials, goods or workmanship are to be to the 'reasonable satisfaction' of the architect/contract administrator.
[470] Burke (2002), 237.
[471] APM (1998), Schedule of Services, clauses 4.1, 4.4, 4.6 and 6.3.

The RIBA Form of Appointment for a Project Manager (RIBA PM (2004)) reflected in its Services Supplement the equivalent standard RIBA work stages of its architect's appointment[472], including the deferral of main contractor appointment until after Work Stages A through to H have been completed, namely until commencement of the construction phase of the project. Nevertheless RIBA PM (2004) included as part of the Project Manager's Services: 'Development and maintenance of a project strategy'[473], which presumably allows the project manager to develop a strategy that includes early contractor appointment.

The creation of a procurement strategy will require the project manager to establish processes for design development, for selection of the design consultants, other consultants, main contractor, subcontractors and suppliers, for pricing of their services and works and for a range of risk management activities. It has been suggested that project managers need to undertake or propose appropriate actions to 'eliminate risks before they occur', or at least to reduce their effect if they are unavoidable, to the extent that such elimination or reduction is 'possible and cost effective'[474]. Adoption of a two-stage procurement strategy, with a supporting conditional preconstruction phase contractor appointment, is one such action that project managers and their clients can adopt if a choice is made early enough in the risk management process.

Lock linked the instigation by the project manager of corrective measures to resolve a problem with a system that ensures adequate warning of the problem. He suggested that this can be achieved by the project manager having a 'well prepared schedule' in place which it keeps up to date and uses to monitor progress on a regular basis[475]. Yet the management of time depends also on all parties trying to keep ahead of the game and acting quickly to mitigate the effects of problems when they arise. It demands an active communication system agreed among all project team members, with a pre-agreed set of key dates to identify any delay in agreed outputs plus an agreed procedure and forum for notifying and reviewing problems under which the parties know to whom warnings should be given and how they will be acted upon[476].

Lock also pointed to 'regular progress meetings' as a project management tactic which can pre-empt problems so that they are prevented

---

[472] RIBA (2004).
[473] RIBA PM (2004), Schedule 2 Services Supplement.
[474] Smith *et al.* (2006), 2.
[475] Lock (2000), 495. See also Section 5.3.6 (Contractual duty to warn of problems).
[476] As considered in Chapter 5, Section 5.3 (The role of communication systems) and Section 5.4 (The role of binding programmes).

rather than needing to be resolved at a later date[477]. However, this is not a convincing preventive method if problems arise in between those meetings or in respect of problems that are not notified because the parties remain silent in order to protect their commercial positions. Again, a more thorough contractual communication system is required, including provision for early warning of problems and an obligation of all team members to seek solutions in response to such early warnings.

### 8.2.3 The role of the project manager in integrating other team members

If the project manager is the party most likely to define the combination of team members and the structure of the team, then logically it is also the party best placed to integrate the roles and responsibilities of those team members. This is reflected to varying extents in the APM and RIBA forms of appointment of a project manager. APM (1998) required the project manager to 'Endeavour to engender a culture of confidence, trust, safe working and mutual respect between all members of the Project Team'[478]. RIBA PM (2004) required the client and its project manager to 'work together in a spirit of mutual trust and cooperation' but does not require the project manager to extend this spirit to the remaining team members[479]. It also included in the project manager's services 'Development and maintenance of a management structure and communications environment in which all consultants, contractors and other persons can perform effectively'[480], but makes no mention here of the client. Where the project manager is appointed as 'Design leader', the RIBA PM (2004) services included 'establishing the form and content of design outputs, their interfaces and a verification procedure'[481]. Where the project manager is also appointed as 'Lead Consultant', such services extended to 'coordinating and reviewing the progress of design work ... facilitating communications between the Client and the Consultants' and 'advising on the need for and the scope of services by Consultants, specialists, subcontractors or suppliers'[482].

The project manager also has a positive role to play in drawing the team together, and its authority to call and chair meetings, combined with its authority to issue instructions under the building contract, puts it in a central position. The contractual role of the project manager can

---

[477] Lock (2000), 507/509.
[478] APM (1998), clause 7.5.
[479] RIBA PM (2004), clause 1.6.
[480] RIBA PM (2004), Schedule 2 Services Supplement B.
[481] RIBA PM (2004), Schedule 2 Services Supplement C.
[482] RIBA PM (2004), Schedule 2 Services Supplement C.

be as a representative of the client or as an independent party acting objectively as between the client and the main contractor or a combination of the two[483]. However, if and to the extent that the project manager's contractual role is as a representative of the client, or if any other aspect of its role compromises or negates the requirement for the project manager to be fair and impartial, a question must arise as to the credibility of the project manager as a unifying influence. If, for example, JCT CA is used for the appointment of a project manager (referred to as 'contract administrator'), there is an express requirement to 'exercise his powers, duties and discretions fairly, impartially and as required by the Building Contract'[484].

Eggleston observed that 'There is no express requirement in NEC3 for the project manager to be impartial'. However, in *Costain* v. *Bechtel*[485] it was determined that, even in the absence of such a requirement under NEC2, there is an implied legal duty of the project manager to act fairly and impartially when certifying payment[486]. A means of demonstrating the objectivity of the project manager is through shared information available to all team members set out in or developed pursuant to a preconstruction phase agreement.

> Project X (newbuild housing): the project manager had a contractual duty to be impartial in its assessment of contractor claims for additional time and money, and a preconstruction phase agreement had been used to build up an agreed timetable for all design releases during the construction phase of the project. However, the project architect, who was also appointed as project manager, undermined its objectivity in the eyes of other team members when it was required to assess main contractor claims for delay that included claims for delay in the release of design details and claims relating to the issue of what it called 'architect's instructions' (not a defined term in the relevant contract) that might or might not have constituted variations with cost and time implications[487].

## 8.2.4 The use of programmes by project managers

A recognised technique for identifying and managing any omission or delay is for the project manager to create a programme, with particular

---

[483] The role is representative under JCT 2005 Design and Build, independent under PPC2000 and a combination of the two under JCT 2005 SBC/WQ.
[484] JCT CA, clause 3.2.1.
[485] *Costain Limited* v. *Bechtel Limited* (2005) EWHC 1018.
[486] Eggleston (2006), 88.
[487] Project case study 8, Appendix A.

activities and deadlines agreed by each project team member, during both the preconstruction phase and the construction phase of a project. A tool extensively used by project managers is the bar chart as originally formulated by Henry Gantt (1861–1919), an American industrial engineer. Lock referred to them as 'day to day control tools' and 'very valuable planning aids'[488]. Burke also described bar charts as providing 'an excellent management tool' that the project manager can use for planning and control purposes[489].

However, neither Burke nor Lock recognised the limited value of bar charts if they are not contractually binding. Progress required by non-binding bar charts can be frustrated by Lock's 'set of excuses' used to evade responsibility at progress meetings[490]. While bar charts will clearly assist in planning, it is questionable as to whether they assist in control unless they have contractual effect requiring adherence to agreed key dates, for example in respect of delivery of consultant designs, procurement by the contractor of remaining prices and proposals, and grant by the client of required approvals.

The use of a contractual programme establishing binding key dates appears to be supported by Lock, who stated that it is important to ensure that 'all significant stages of the project must take place no later than their specified dates'[491]. These stages logically include the preconstruction stages. Lock also recognised that failure to start a particular stage of work on time is a common project risk that can result from delays caused not only by third party events, but also from problems within the team such as procrastination, shortage of information or lack of resources[492]. The question is how the parties can head off these problems before they arise.

The majority of the causes of delay cited by Lock (procrastination, shortage of information, lack of resources) could be identified and averted or limited by means of a contractual programme governing key dates for relevant preconstruction phase activities. This would allow the project manager a greater ability to exercise the measures of control that are among the stated purposes of project management. APM (1998) recognised the importance of the programming obligations of the project manager at each stage of the project, but does not state whether the programmes created should have contractual effect or how the project manager should exercise any authority over other team members if programmes do not have contractual effect.

---

[488] Lock (2000), 168.
[489] Burke (2002), 154.
[490] Lock (2000), 510. See also Chapter 5, Section 5.3.2 (Notices and meetings).
[491] Lock (2000), 7.
[492] Lock (2000), 7.

APM (1998) envisaged that the project manager will prepare 'a programme for the Project' which will be signed by the client and the project manager (but why not the other team members?) and will be subject to revision by the project manager, who must use reasonable skill, care and diligence to perform its services 'at such time or times as shall be appropriate having regard to the Programme'[493]. In relation to the project manager's programming obligations, APM (1998) included in its services the following:

- 'Initiate action in the event that any aspect of the Project appears to be likely to fail to achieve the Client's objectives, public obligations, budget and programme. Agree suitable corrective action and monitor its implementation.
- Develop a plan for execution of the Project through all stages to its handover.
- Prepare an outline Programme and preliminary Cost Plan.
- Initiate the preparation of a detailed implementation programme'[494].

The RIBA PM (2004) form provided for the following services:

- 'Development, implementation and maintenance of project procedures;
- Preparation and maintenance of a master programme, coordination with any programmes prepared by Consultants or contractors'[495].

JCT CA, in relation to the role of the project manager as contract administrator, requires release of design information and other information in accordance with any agreed 'Information Release Schedule' forming part of the Building Contract[496], although this will not cover programming of the preconstruction phase of the project.

APM (1998), RIBA (PM) 2004 and JCT CA all contain some programming provisions and these can be connected to the agreed procurement strategy. However, in each case they will benefit from greater detail as to the contractual effect of the programmes created and the authority commanded by the project manager in securing adherence to those programmes by other consultants and by contractors. These details are made clear in the role of the Client Representative as project manager under PPC2000 and its authority to issue instructions in relation to

---

[493] APM (1998), clause 2.4.
[494] APM (1998), Model Schedule of Services, clauses 1.2, 3.7, 4.8 and 7.3.
[495] RIBA PM (2004), Schedule 2 Services supplement, B.
[496] JCT CA, clause 3.2.2.

compliance by all team members with the Partnering Timetable and Project Timetable[497]. They are also clear in the authority of the project manager under NEC3 and in the NEC Professional Services Contract as regards the requirement for compliance with a Completion Date and agreed Key Dates[498].

## 8.2.5 The use of preconstruction phase agreements by project managers

The effectiveness of a project manager will depend significantly on personal influence over other team members. A confident and assertive project manager who understands clearly the roles and responsibilities of other team members can add considerable value and may take the view that it does not need additional contractual mechanisms to support it in doing its job.

If a project manager has power as the client's appointed intermediary to deal with other project team members, then the exercise of that power may be influential on the project, irrespective of supporting contractual processes. However, the project manager's confidence and power can also have an adverse effect on the project if full knowledge of the way that the respective roles and responsibilities of project team members fit together, and their interrelated timelines, resides only with the project manager[499].

A project manager who wishes to facilitate efficient project processes, and to have the benefit of a communication system and programme of agreed activities at all stages, should welcome the availability of a preconstruction phase agreement, not least as a means to ensure that all team members have agreed to a consistent approach. The clearer the agreement governing preconstruction activities and programmes, the easier it should be for the project manager to do its job, namely to foresee, predict, plan, organise and control project activities by means that are preventative rather than curative. The need for a proactive approach was emphasised by Lock, who stated that 'Any project manager worthy of the title will want to make certain that whenever possible his or her tactics are preventative rather than curative'[500].

---

[497] PPC2000, clauses 5.1, 5.3, 5.5, 6.1 and 6.5.
[498] NEC3, clauses 25.3 and 30 and NEC Professional Services Contract, clause 30.
[499] See also Chapter 3, Section 3.8 (Links between claims and building contracts) for Judge Anthony Thornton's observations regarding lack of information sharing between project participants and Chapter 4, Section 4.3.3 (Information required for accurate pricing) for Milgrom & Roberts' observations as to private information creating inefficient and unbalanced relationships.
[500] Lock (2000), 507.

## 8.3 Preconstruction phase agreements and partnering

### 8.3.1 What is partnering?

Many definitions have been provided for partnering, and this has been the cause of much confusion in the construction industry. For example, partnering can be presented as a fuzzy set of relationships dependent on altruistic behaviour and the voluntary compromise of commercial rights and obligations. It can therefore appear fragile or even imprudent when held up to a harsh commercial light.

A definition offered by Barlow *et al.* was that 'Partnering is simply a generic term for a range of practices designed to promote greater co-operation between organisations'. Barlow *et al.* also observed that 'many features of partnering – especially its emphasis on the management of people across organisational boundaries – are not new to the construction industry'[501]. They recognised that 'In the construction industry a degree of collaboration has always been common and may be the norm among the specialist trades and small builders'[502]. However, Barlow *et al.* believed that partnering can be more closely defined, in that 'What distinguishes relationships involving partnering from other forms of alliance is their emphasis on improving the performance of each party'[503].

To translate the generic concept of partnering into working practices requires a clear understanding of the ways that different businesses interact when formulating a procurement strategy, assuming responsibilities, calculating rewards, agreeing information systems, undertaking joint working and allocating appropriate resources. Smith *et al.* saw partnering as a type of project management and described it as 'a structured management approach to facilitate teamworking across contractual boundaries'[504]. The National Economic Development Council noted that, whatever the relationships established and techniques used, 'the main objective of partnering is to meet the client's requirements in the most cost-effective way'[505]. Barlow *et al.* observed that 'partnering is a process rather than a particular form of relationship between organisations'[506], and this gets us to the heart of the issue. If partnering is a process rather than just a relationship, then that process should be capable of description and needs to be described so that all team members can understand and follow it in the same way.

---

[501] Barlow *et al.* (1997), 1.
[502] Barlow *et al.* (1997), 58.
[503] Barlow *et al.* (1997), 58.
[504] Smith *et al.* (2006), 144.
[505] NEDC (1991), 9.
[506] Barlow *et al.* (1997), 58.

*Project Management and Project Partnering*

What type of relationship or process qualifies as partnering? Is it one characterised by teamwork? The RIBA Guide states that 'In a partnering relationship all the contributors (partners) including the client, agree goals for the success of the project and will benefit from better performance'[507].

Banwell in 1964 recognised the benefits of 'co-operation in design and construction'[508], and that 'The contractor appointed at an early stage will be able to develop a close relationship with all the other partners in the design and construction team before work begins on site'[509]. CIRIA noted in 1998 that 'There is increasing recognition that the interests of all parties are served best by adopting genuine teamwork' and observed that 'A positive and collaborative attitude on the part of the project team can deliver significant benefits during the execution of any project, and is an essential requirement if contractors are to add any value beyond simply completing work in accordance with the contract'[510]. However, CIRIA emphasised that while recognition of the benefits of teamwork has in part encouraged partnering and alliancing, teamwork itself is not confined to partnering arrangements. If teamwork is required for partnering, but is not exclusive to partnering, then is it preferable to see partnering less as a relationship than as a set of processes that give rise to that relationship?

The links between project management and team-building that are required for successful project partnering are termed 'action centred leadership' by Burke. This approach combines the following:

- The needs of the individuals, which need to be supported by personal motivation;
- The needs of the team, which need to be supported by integration and team-building;
- The needs of the project, which need to be supported by planning and control.

Burke emphasised the need 'to deliver the project objectives; scope, time, cost and quality through an effective planning and control system which includes integrating the project process, communication, organisation structures and risk management'[511]. He suggested that such a system creates an environment for the team to solve problems and to make good decisions that may determine the success of the project[512].

---

[507] RIBA Guide, 45.
[508] Banwell (1964), 5, Section 2.7.
[509] Banwell (1964), 10, Section 3.15.
[510] CIRIA (1998), 21.
[511] Burke (2002), 279.
[512] See Burke (2002), 279. See also Chapter 2, Section 2.3 (Effect of the number of parties) regarding the challenges of motivation.

*Early Contractor Involvement in Building Procurement*

However, to focus excessively on the collaborative relationships created through team-building can be to the detriment of effective project planning and control techniques.

> Watergate School, Lewisham (newbuild school): the client, design consultants and main contractor focused their efforts on working together to produce the best possible designs, but started to neglect the design deadlines in the preconstruction phase programme for the pricing of those designs as supply chain packages so as to achieve timely start on site[513].

As large numbers of construction partnering projects are implemented, and as the industry and its clients seek an explanation of what this approach involves, there is increasing recognition that partnering is primarily a planning system providing a way, in the words of Bennett & Pearce, to 'lay the foundations for meeting agreed objectives by planning the design, construction and completion of each major stage of [their] work as an integrated system'[514].

### 8.3.2 Types of partnering and their potential benefits

There are perceived to be two primary types of partnering, one confined to the implementation of a single project ('project partnering') and the other extending to a series of projects ('long term partnering'). Barlow *et al.* described all partnering as 'a set of collaborative processes'[515], but distinguished between project partnering and long-term partnering as follows:

Project partnering:

- 'Generally refers to a much narrower range of co-operative arrangements between organisations for the duration of a specific project'[516].

Long-term partnering:

- 'Covers a broad range of strategic cooperative relationships between organizations ... and can involve highly structured agreements providing for a high level of cooperation between partners'[517].

---

[513] Project case study 2, Appendix A.
[514] Bennett & Pearce (2006), xi.
[515] Barlow *et al.* (1997), 6.
[516] Barlow *et al.* (1997), 7.
[517] Barlow *et al.* (1997), 6.

*Project Management and Project Partnering*

An NEDC report in 1991, which focused on long-term partnering, found that 'the prime client motive behind the establishment of partnering arrangements is the client's wish to reduce his overheads and give greater concentration to his core business'. It concluded that from this primary motive flow benefits by way of shared expertise and information as well as reduced tendering costs and the development of total quality management, improved safety procedures, improved training and resource development, and the opportunity for innovation[518].

Long-term partnering is frequently governed by framework agreements which are considered in Chapter 7. This chapter examines partnering primarily in the context of project partnering in order to explore the effect of partnering as a project management approach on the treatment of preconstruction phase processes.

As to the potential benefits to be derived from partnering, the Ministry of Defence in their Partnering Handbook stated that 'The only justification for partnering is the pursuit of better value for money which needs to be continually demonstrated'[519]. Barlow *et al.* observed that these can include significant improvements in productivity, cost savings and improved innovation[520]. In addition, construction clients have reported:

- Substantial reduction of construction costs and delivery time[521];
- Improvement of construction quality[522];
- Intelligent resolution of problems and far fewer disputes[523].

Equivalent benefits were identified by the National Audit Office. They identified 'value for money gains' from 'partnering and collaborative working' including 'streamlined procurement ... leading to improved productivity', 'reduced construction costs', 'improved whole life value', 'reduced legal claims' and 'improved health and safety'[524].

However, the economic cycle and other influences unrelated to partnering can give rise to misleading success stories and it is important to consider the wider context when examining the claimed benefits of partnering. Barlow *et al.* recognised 'Non-partnering influences on performance' such as the ability of clients to obtain price reductions from contractors and suppliers at a time of construction industry recession,

---

[518] NEDC (1991), 1.
[519] MoD (2007), 5.
[520] Barlow *et al.* (1997), 44.
[521] See for example Project case studies 5.6 and 7.
[522] See for example Project case studies 2, 3 and 4.
[523] See for example Project case studies 1, 4 and 6.
[524] NAO (2005), 43.

and also the benefits that can result from general improvements in construction technology[525].

### 8.3.3 Commercial cooperation and behavioural change

Bennett *et al.* saw partnering as an evolving phenomenon and recognised that a system of cooperative decision making could use feedback from earlier projects to improve performance on later projects. They saw partnering evolving through 'three generations' of increased integration. In the first generation, mutual objectives, decision-making and continuous improvement are all structured with recognition of partnering, but without new project processes.

In the second generation of partnering, new project processes are put in place around which the following revolve:

- Strategy – developing the client's objectives and how consultants, contractors and specialists can meet them on the basis of feedback;
- Membership – identifying the firms that will need to be involved to ensure all necessary skills are developed and available;
- Equity – ensuring everyone is rewarded for their work on the basis of fair prices and fair profits;
- Integration – improving the way the firms involved work together by using cooperation and building trust;
- Benchmarks – setting measured targets that lead to continuous improvement in performance from project to project;
- Project processes – establishing standards and procedures that embody best practice based on process engineering;
- Feedback – capturing lessons learned from projects to guide the development of strategy'[526].

In the third generation, further integration between construction firms and their regular clients uses cooperation throughout their supply chains to build up 'virtual organisations that respond to and shape rapidly changing markets'[527].

Bennett *et al.* identified new 'project processes' as the basis for 'second generation' partnering[528]. These are consistent with the creation of such project processes early in the life of the project, for example as a means of 'developing the client's objectives', identifying all team members 'to

---

[525] Barlow *et al.* (1997), 57. See also Chapter 10, Section 10.11 (Preconstruction phase agreements in an economic downturn).
[526] Bennett & Jayes (1998), 4.
[527] Bennett & Jayes (1998), 5.
[528] Bennett & Jayes (1998), 4.

ensure all necessary skills are developed and available' and 'establishing standards and procedures that embody best practice based on process engineering'[529].

In order to achieve the reasonableness and fair dealing among team members that is a claimed feature of partnering relationships, certain commentators have proposed that it is necessary for organisations to achieve cultural alignment whereby they share the same values, attitudes and beliefs[530]. However, Bresnen & Marshall queried whether changing behaviour in fact depends upon changing deeper underlying attitudes, beliefs and values. They suggest that instead it is possible that compliance with agreed partnering arrangements (for example contractual partnering arrangements) can be achieved for commercial reasons that leave the different organisational cultures intact. Bresnen & Marshall questioned 'whether in fact it is possible to manipulate and change organizational culture in the ways commonly prescribed' in relation to partnering, and also doubt whether it is possible 'to standardize models of partnering "best practice"' [531]. In their view, the 'technical appariti' of partnering (which they recognise includes contracts and pricing formulae as well as charters and workshops) may also be insufficient to establish successful partnering relationships, even if supported by organisational alignment in the absence of sound commercial justification[532].

This is a refreshing approach that questions the behavioural idealism of some partnering enthusiasts and recognises that successful collaboration may be achieved without necessarily changing the nature of the collaborators if there is sufficient clarity as to what actions are required and what benefits they will achieve. To recognise that 'business is business' irrespective of collaborative relationships will strike a chord with many clients and contractors and should help dismiss once and for all the idea of partnering as standing separate from normal commercial rules of engagement. This approach to partnering is supported by the Ministry of Defence who state 'The existence of a partnering contract does not mean that customer and supplier have all interests in common or that the supplier will not seek, quite reasonably, to optimise their commercial interests or commercial return within the parameters of our partnering relationship with them'[533].

Partnering as a system of agreed actions and rewards, rather than only as a set of shared values, immediately makes it more accessible as a way of doing business. Developing this theme, it has also been

---

[529] Bennett & Jayes (1998), 4.
[530] See, for example, Bresnen & Marshall (2002), 234.
[531] Bresnen & Marshall (2002), 235.
[532] Bresnen & Marshall (2002), 235.
[533] MoD (2007), 3.

suggested that, for the real benefits of partnering to be achieved, it needs to be customised to reflect the particular needs of the parties and the project to which it is applied. In Bresnen & Marshall's view, this requires, in addition to the 'technical appariti', an honest assessment of whether the changes induced in the relevant project circumstances are deep enough to be changes in attitude or whether they simply reflect behavioural compliance based upon calculations of self-interest[534].

The success of such customised partnering demands consistency and clarity in contracts to set out the agreed partnering actions and rewards, as the parties cannot assume that for any non-commercial reason the other team members will change their behaviour or their organisational structure in order to subscribe to a set of partnering values.

### 8.3.4 Challenges to successful partnering

Partnering is attractive to many parties because it appears to make the process of project procurement more easily understood. Client organisations in particular, where they lack detailed technical and financial knowledge relevant to design, supply and construction activities, may be drawn to partnering relationships and techniques that are expressed in cultural and behavioural language. The consequent risk is that such parties may underestimate the complexity of the building project and may pay insufficient attention to the views of those who are involved in the complex mechanisms of its implementation. Jones *et al.* observed that:

> 'Collaborative working requires careful pre-planning and close management of all the various inputs (design, manufacturing know-how, installation expertise, cost and value management, hands-on project management, explicit risk management and so forth). Parties entering into partnering arrangements who focus simply on cultural behaviour changes rather than managing risks may find that rose-tinted glasses obscure their vision. When some of their aspirations start to unravel because the risks that impact time, cost or quality issues have not been addressed in a robust manner, the project and participants suffer'[535].

This risk can only be addressed through detailed contractual arrangements that are clear as to all parties' commercial rights and obligations in relation to every aspect of the project at each stage of its progress.

---

[534] See Bresnen & Marshall (2002), 235.
[535] Jones *et al.* (2003), 189.

There are a number of other risks to successful partnering, which N.J. Smith described as the 'four major road blocks to success'. The four roadblocks are:

- 'A shift in business conditions: if conditions change and the project is behind schedule, with unanticipated technical problems and cost overruns, the strategy within each organisation may change and even revert to an "us versus them" attitude.
- Uneven levels of commitment: unevenness of commitment often develops from the basic differences between organisations.
- Lack of momentum: a partnership requires nurturing and development throughout the life of the project. The representatives from each side must constantly work to maintain the health of the partnership.
- Failure to share information: partnering requires timely communication of information and the maintenance of open and direct lines of communication among all members of the partnering team. The failure to share information is most likely to arise when team members revert to past practices'[536].

It is arguable that each of these road-blocks can be steered around with the support of clear provisions in a building contract:

- A shift in business conditions: this is not necessarily a reason to revert to an adversarial position if there is a contractual entitlement to receive open information as to cost so that the implications of the shift can be examined by all parties[537]. Such a shift can also be managed if there is a medium under the building contract (e.g. a core group) through which team members can share a common understanding of any consequent problem (even if it is the contractual responsibility of only one party) and can seek an alternative solution that is of benefit to the project[538]. However, this approach is likely to be more successful if there is a long-term framework relationship to justify the parties compromising short-term self-interest[539].
- Uneven levels of commitment: these can be tested and ironed out during the preconstruction phase if the project team members sign up to an early agreement clarifying their respective commitments

---

[536] Smith N.J. (2002), 304.
[537] See Chapter 4, Section 4.3 (Preconstruction pricing processes).
[538] See Chapter 5, Section 5.3 (Role of communication systems).
[539] See Chapter 7 (Preconstruction commitments under framework agreements), but see also Chapter 9, Section 9.2.4 (Single-stage tendering) as to the concerns of clients in respect of changing business conditions that affect prices.

and the ways in which they expect to rely on each other's commitments (e.g. meeting deadlines for design release)[540].
- Lack of momentum: this can be addressed through commitment to key dates for completion of key activities of all team members during the preconstruction phase and construction phase[541]. Avoiding delays is dependent on leadership of the agreed partnering processes by means of the project manager requiring adherence to deadlines, and also by means of a medium (e.g. a core group) through which representatives of each party can exercise peer group pressure in the event of any delay or other under-performance.
- Failure to share information: this can be addressed through the early establishment of a communication system that requires the parties to share required information, including open-book pricing to build up detailed costs[542]. Such a system should require the use of clearly agreed channels for communication and the appointment of consistent representatives of each party, operating within agreed terms of reference (e.g. a core group).

### 8.3.5 Partnering and building contracts

A perceived problem in developing an understanding of partnering is the limited capacity for rapid change through organisational learning. Barlow *et al.* remarked that it is difficult for an organisation working on its own to learn the means by which to operate in a different way, and it is therefore much harder for a number of organisations to achieve significant change when working together[543]. Bresnen & Marshall went further than this and questioned whether it is possible to 'engineer' short-term collaborative processes in the manner that partnering is alleged to achieve[544]. Where mutual objectives, trust and an understanding of each other's commitments need to be developed, they observe that, although it is widely recognised that attitudinal and behavioural issues are important to successful partnering, there is less clarity as to the means by which these characteristics are to be created or encouraged.

Forms of professional appointment and building contract, including conditional preconstruction phase agreements, have an important role

---

[540] See Chapter 6 (Contractual and non-contractual options) and also Chapter 5, Section 5.4.2 (Preconstruction phase programmes), but see also Chapter 9, Section 9.4 (Personal obstacles) as to varying personal attitudes.
[541] See Chapter 5, Section 5.4 (The role of binding programmes).
[542] See Chapter 5, Section 5.3 (The role of communication systems) and also Chapter 4, Section 4.3 (Preconstruction pricing processes).
[543] Barlow *et al.* (1997), 7.
[544] Bresnen & Marshall (2002), 232.

to play in the achievement of successful project partnering. Rhys-Jones noted that 'The encouragement of co-operative attitudes is greatly influenced by the contractual arrangements'[545]. She cited the American Society of Civil Engineers' document *Quality and the Constructed Project*, which stated that 'Just as the design professional is employed to prepare a workable set of drawings and specifications, the attorney should be employed to review and approve a workable set of contracts that address proper business relationships, communication channels and quality control issues'[546].

Although N.J. Smith did not see contracts as 'an important part of the partnering process', he recognised that his view reflected 'a purist standpoint' and accepted that a contract should promote and complement the partnering processes because otherwise it would undermine them and contribute to adversarial behaviour[547]. The apparent contradiction in this statement reflects the wistful notion, still surprisingly widespread in the industry, that a contractual halfway house should be created, where partnering can be installed, separate from the harsh practicalities of contract administration[548].

The RIBA suggested that there are two alternative ways to document partnering whereby:

> 'The relationships, the objectives and the procedures are drawn up with legal assistance and recorded:
> 
> - In a project-specific "charter" – a supplement to the separate agreements between client and each of the partners, e.g. SFA/99 for the architect; or
> - In a partnering agreement signed by all of the partners'[549].

Without full integration in the building contract and consultant appointments, there is the risk that partnering will be treated as an optional set of management processes as there will always be a separate adversarial contract to fall back on. This creates a number of problems, as illustrated in the 2000 Housing Forum Survey of organisations involved in partnering: 'In some cases, there was a lack of understanding of the partnering arrangement by some partners and, more often, by those not included in it. This lack of understanding could cause confusion between partnering arrangements and negotiated

---

[545] Rhys-Jones (1994), 6.
[546] Rhys-Jones (1994), 3.
[547] Smith N.J. (2002), 299, 300.
[548] See also Chapter 6, Section 6.4 (Non-binding agreements).
[549] RIBA Guide, 45.

contracts. It also led to a lack of involvement of suppliers and users'[550]. This survey result shows growing industry demand for clarity as to exactly where partnering sits in the contractual relationship.

The successful establishment of partnering takes place during the preconstruction project planning phase and is therefore dependent not only on its integration in the construction phase building contract, but also on its treatment in a preconstruction phase agreement. The NEDC appeared to infer the need for a preconstruction phase partnering agreement when they stated that: 'The partnering agreement has led to both sides working together earlier than usual during the design phase'[551].

The CIC suggested in 2000 that 'An effective partnering contract should support the full partnering team and aim to deliver an integrated project process. Logically, it should replace any of the existing standard forms'. The CIC went on to propose that 'An effective contract can play a central role in partnering. It sets out the common and agreed rules; it helps define the goals and how to achieve them; it states the agreed mechanism for managing the risks and the rewards; it lays down the guidelines for resolving disputes'[552].

NEC2 in 1995 and GC/Works/1 in 1999 had already included wording designed to encourage a cooperative approach, as quoted below. Following publication of the CIC recommendations, a number of published other standard form contracts have included express project partnering provisions, namely:

- PPC2000, published in 2000 and amended in 2003;
- NEC Partnering Option X12, published in 2001 and incorporated in NEC3 in 2005;
- Perform 21 Public Sector Partnering Contract, published in 2004;
- JCT 2005 Framework Agreement, published in 2005;
- JCT Constructing Excellence Contract, published 2006.

The extent to which each of these standard form contracts deals with preconstruction phase processes, programmes and communication systems is discussed elsewhere[553]. What they have in common is a commitment to partnering that goes beyond non-contractual (and therefore arguably optional) arrangements and seeks to influence the behaviour of the parties by means of contractual terms.

The difficulty of defining partnering values in a contractual format is illustrated by the general wording used in standard form building

---

[550] Housing Forum (2000), 13.
[551] NEDC (1991), 78.
[552] CIC (2002), 12.
[553] In Chapters 4, 5 and 6 and Appendix B.

contracts. Each of the following statements promotes project partnering:

- GC/Works/1: 'The Employer and the Contractor shall deal fairly, in good faith and in mutual co-operation, with one another, and the Contractor shall deal fairly, in good faith and in mutual co-operation, with all his subcontractors and suppliers'[554].
- NEC 2 and NEC3: 'The Employer, the Contractor, the Project Manager and the Supervisor shall act as stated in this contract and in a spirit of mutual trust and co-operation'[555].
- PPC2000: 'The Partnering Team members shall work together and individually in the spirit of trust, fairness and mutual cooperation for the benefit of the Project, within the scope of their agreed roles, expertise and responsibilities as stated in the Partnering Documents' [556].
- Perform 21 Partnering Agreement: 'All Partners will deal fairly with each other and work together in a spirit of mutual trust, good faith and cooperation and apply their agreed expertise in relation to the project'[557].
- JCT 2007 Framework Agreement: 'The Parties will continually impress upon all personnel involved with the Tasks their keen desire to work with each other and with all other Project Participants in an open, co-operative and collaborative manner and in a spirit of mutual trust and respect with a view to achieving the Framework Objectives'[558].
- JCT CE: Under an 'overriding principle' the parties agree to 'work together with each other and with all other Project Participants in a co-operative and collaborative manner in good faith and in the spirit of mutual trust and respect'[559].

However, the effectiveness of general contract wording that promotes partnering values is tested by the extent to which other contractual provisions provide the means by which project team members are obliged or encouraged or allowed to apply their agreed partnering values in practice, and this is examined in relation to each of the above standard forms in Appendix B.

The wisdom of relying on general statements of partnering values is now questionable. Judge Humphrey Lloyd in *Birse Construction* v.

---

[554] GC/Works/1, clause 1A of Conditions of Contract.
[555] NEC2 and NEC3, core clause 10.1.
[556] PPC2000 Partnering Terms, clause 1.3.
[557] Perform 21 PSPCP (2004), clause 2.
[558] JCT 2007 Framework Agreement, clause 9.1. See also Appendix C.
[559] JCT CE, clause 2.1.

St David recognised the contractor's expectation that 'the partnering ethos ... would naturally have led to a sympathetic approach to questions of extension of time and deduction of damages for delay'[560]. This lack of certainty can undermine clearly agreed contractual terms and lead to unpredictable results in the event of a dispute. For this reason, the provision in JCT CE for an 'overriding principle' should be a cause for concern as it may have the effect of qualifying the enforceability of all other JCT CE contractual terms. A contract that fully integrates partnering values with other contractual provisions should provide a more stable basis for successful partnering relationships.

### 8.3.6 Fit of partnering and contractual relationships

Notwithstanding the existence and use of both preconstruction phase agreements and construction phase building contracts that support partnering relationships, questions continue to arise regarding the fit between partnering values and contracts. Four instances were raised by Skeggs where in his view contracts might be compromised by partnering in a common law jurisdiction:

(1) Good faith and ambiguity as to 'whether there is an implicit duty to perform a construction contract in good faith in common law jurisdictions'[561];
(2) Estoppel and waiver if 'in a partnering arrangement parties may make representations to one another which do not conform to the contract but upon which they rely'[562];
(3) Confidentiality and statements made 'without prejudice' if 'successful partnering requires a degree of disclosure which could compromise a party's position on the project or outside of the project environment'[563];
(4) Fiduciary relations where 'the participants in a partnering agreement must consider whether they owe fiduciary obligations to the other partnering parties which impinge upon their right to act in their own self-interest'[564].

Each of these questions are directly relevant to the enforceability of those contracts that recognise partnering as a means of implementing preconstruction phase processes.

---

[560] *Birse Construction Limited* v. *St David Limited* (2000) 1BLR57.
[561] Skeggs (2003), 463.
[562] Skeggs (2003), 463.
[563] Skeggs (2003), 464.
[564] Skeggs (2003), 464.

### 8.3.7 Good faith and fiduciary relations

Dealing with Skeggs' first and fourth points, both PPC2000 and NEC3 and the JCT 2007 Framework Agreement steer clear of the words 'good faith'. GC/Works/1[565] and Perform 21[566] expressly provide for an obligation to deal 'in good faith', but do not expand on how this affects the other contractual duties of the project team members.

JCT CE offers a three-track approach which risks confusion and whereby:

- The parties agree to a basic set of contractual obligations.
- These basic contractual obligations are subject to the 'overriding principle' which includes 'good faith' and which is to be taken into account in the event of any dispute, although it is not stated by what rules the 'overriding principle' will be applied to amend basic contractual obligations[567].
- The parties also conclude a 'project protocol'[568] describing their objectives in their own words, but without creating any additional contractual obligations. This raises the question of why a construction contract should encourage the project team members to work on an additional non-contractual document which they may wish to use (as they created it in their own words) but which will have no legal effect, so that if the parties at any stage require legal and contractual clarity they will have to abandon this document and revert to the building contract.

Although it does not use the words 'good faith', it would seem that the JCT CE contract is vulnerable to the arguments that Skeggs puts forward regarding ambiguity in relation to required performance.

### 8.3.8 Estoppel and waiver

As to Skeggs' second point, it is not clear where he believes representations would appear in a partnering arrangement that do not conform to the building contract, but it is assumed that this comment is directed towards the possible contents of a partnering charter. Having noted above the risk of inconsistencies between the JCT CE contract and its overriding principle and its project protocol, the same risk of representations that do not conform to the underlying building contract could

---

[565] For example, GC/Works/1, clause 1A(1) (Fair dealing and teamworking).
[566] Perform 21 PSPCP, clause 2.
[567] JCT CE, clauses 2.1 and 2.9.
[568] JCT CE, clause 1.4.17.

be raised in respect of the partnering charter published by the JCT and in respect of the JCT 2007 Framework Agreement, when either are read in conjunction with JCT 2005 building contracts[569].

The CIC Guide[570] and PPC2000[571] recognise that the project team members may utilise a partnering charter as a means to summarise their aims and objectives. However, the CIC Guide notes that such a partnering charter must not be inconsistent with the terms of the underlying building contract[572], and PPC2000 states that it should form one of the contract documents with a stated place in the order of priority of those contract documents[573].

### 8.3.9 Confidentiality and disclosure

As to Skeggs' third point, this is linked to the contractual duty to warn that appears, for example, in NEC3 and PPC2000[574]. It is possible that such a duty to warn could clash with the requirements of a project team member's professional indemnity insurance policy and could invalidate such insurance, which clearly would not be in the interests of the client or the project. A related problem is the insistence of many professional indemnity insurers that their clients do not disclose the terms of their policies to other project team members. As a result, it is often difficult for parties other than the insured to identify any relevant restrictions in the terms of a professional indemnity insurance policy and to ascertain what, if any, level of disclosure is permissible by way of early warning. Professional indemnity insurers remain willing, for the time being, to cover project team members entering into NEC3 and PPC2000 contracts, notwithstanding their contractual duty to warn, so at present the issue is not an obstacle to project partnering in practice.

However, as regards notification of insurance claims themselves, professional indemnity insurers objected to the original wording in PPC2000 that required a project team member who is aware of a claim or potential claim to 'notify the Client Representative of such claim or potential claim and keep the Client Representative regularly informed as to the progress of such claim or potential claim'[575]. That led to an amendment to PPC2000 in 2003 qualifying this obligation so that it applies to each project team member 'subject only to any restrictions

---

[569] See JCT Practice Note 4 and JCT 2007 Framework Agreement.
[570] CIC (2002), 14, Note 5.
[571] PPC2000 Partnering Terms, clauses 2.2(viii) and 2.6(v).
[572] CIC (2002), 14, Note 5.
[573] PPC2000 Partnering Terms, clause 2.6.
[574] PPC2000 Partnering Terms, clauses 2.5 and 3.7, and NEC3 core clause 16.1.
[575] PPC2000 (prior to the 2003 amendments), Appendix 4, Part 4, Item 5.

imposed by its insurers and approved in advance by all other Partnering Team members'[576].

As regards confidentiality obligations generally, commercial confidentiality is not inconsistent with project partnering. For example in PPC2000 the requirement for team members to 'permit inspection of their activities and records in relation to the Project'[577] is subject to an obligation not to reveal to any third party any information 'if and to the extent that it is stated or known ... to be confidential'[578].

This chapter has identified the potential for better recognition of the connections between the responsibilities of project managers and the use of two-stage procurement methods and preconstruction phase agreements. It has also considered the role of partnering as a means to establish cooperative relationships, but has questioned whether such relationships can be commercially robust or even properly understood if they are not fully integrated in contractual preconstruction and construction phase commitments and rewards. It is suggested that contractual commitments to project partnering should be set out alongside project processes in a preconstruction phase agreement and that otherwise such commitments are vulnerable to later misunderstanding, abandonment or unenforceability.

Bennett *et al.* suggested that 'Partnering works by making careful plans at the start of projects and then relentlessly putting them into effect'[579]. This is the role of the preconstruction phase agreement.

---

[576] PPC2000 Appendix 4 Part 4, Item 5.
[577] PPC2000 Partnering Terms, clause 3.11.
[578] PPC2000 Partnering Terms, clause 25.5.
[579] Bennett & Pearce (2006), 83.

# CHAPTER NINE
# OBSTACLES TO EARLY CONTRACTOR APPOINTMENTS

## 9.1  Introduction

Although this book has argued the case for the early appointment of main contractors and certain subcontractors and for the use of preconstruction phase agreements to clarify the early processes to be followed by them in conjunction with other team members, the majority of projects continue to be procured by single-stage selection of contractors under building contracts governing only the construction phase. It is, therefore, important to examine why, despite their demonstrable benefits, early contractor appointments and preconstruction phase agreements are used on only a minority of projects.

This chapter divides possible obstacles to the use of preconstruction phase agreements into those that are specific to a particular type of project, those that derive from procedural and cultural issues, and those that may stem from a lack of awareness of what is possible. It recognises natural limitations and exceptions to the two-stage procurement and contracting approach, but questions whether other reasons given for its rejection are logical or sustainable.

## 9.2  Project-specific obstacles

### 9.2.1  Size and simplicity

The potential for preconstruction phase processes may be restricted by practical considerations. For example, the size or simplicity of the project may not justify close attention to preparatory processes. Similarly, the client may not be willing or able to make available sufficient time for such processes, particularly if the client has no experience of previous projects or if building project processes are outside the client's expertise.

In small or straightforward projects it may be possible for an architect or engineer to complete all design work before inviting tenders

*Obstacles to Early Contractor Appointments*

from prospective main contractors, and for those prospective main contractors to submit prices to take account of all relevant risks[580]. In these cases, any scope for joint working between the consultants and the selected main contractor during the preconstruction phase is likely to be negligible. Where there are preconstruction exchanges between the design consultants and main contractor, their duration may be short and such exchanges will be ad hoc and informal. It is not realistic to expect the parties to formalise their preconstruction phase arrangements in relation to such projects.

## 9.2.2 Design and build projects

In other projects, irrespective of their size, the client may simply wish to state its performance requirements and to take no further part in the project, paying a fixed price to a main contractor for meeting those requirements, whatever price contingencies may be included in that price to take account of deficiencies in the client's brief and programme or of other perceived main contractor risks. Such a client will expect to hand over responsibility for all design procurement and construction activities to a main contractor, and is likely to wish to maintain an arm's length relationship, leaving the main contractor to undertake both preconstruction phase and construction phase activities in isolation.

This approach, known as 'design and build', relies on the client expressing a clear performance brief at the outset and leaving the main contractor to determine its own means to complete the project in line with the required cost, time and quality parameters[581]. The client will have little information regarding progress or problems arising during the life of the project, and may not become involved unless and until such problems escalate to the level of a formal dispute.

Many design and build projects involve some measure of monitoring by a professional project manager on behalf of the client. The ability of the project manager to identify problems in time for them to be rectified will depend not only on its own competence, but also on the quality and detail of the client's brief. The quality of the client's brief, and of its interpretation by the main contractor when pricing the project, may lead to inclusion of significant risk premiums in the price. If, as is often

---

[580] Such projects may be governed by a standard form minor works contract such as JCT 2005 Minor Works or the NEC3 Short Contract, neither of which provide for any preconstruction phase processes.

[581] Design and build projects may be governed by a standard form contract such as JCT 2005 Design and Build. The first JCT design and build form, JCT WCD and its 1998 successor, were extremely popular and as recently as 2004 accounted for 35% of all contracts in use, RICS (2004), 14.

the case in design and build projects, prices at tender stage are not required to be supported by a detailed breakdown, such risk premiums will not be apparent and will be difficult to reduce by negotiation in the absence of an agreed preconstruction process of joint risk management.

An absent client puts considerable pressure on its project manager, as well as its design and build contractor, as regards interpretation of its brief. A further problem arises if the client wishes to make any change to its brief after implementation of the project has commenced, or where claims arising from any event of delay and disruption need to be evaluated. The pricing proposals submitted by the main contractor for a design and build project at the point of award of contract may not be supported by sufficient cost and other details to allow the client or project manager a clear understanding as to how to calculate the cost and time effects of any later change or risk event.

### 9.2.3 Management contracting and construction management

The procurement models known as management contracting and construction management provide for early main contractor appointment for reasons not related to joint working on preconstruction phase processes, but which are instead usually dictated by time constraints and the need for an early start on site. 'Management contracting' involves early appointment of a management contractor to procure a series of works packages, early ones starting on site while others are still being designed. Although the management contractor has the authority of a main contractor over the package contractors, its warranty to the client is limited to amounts recoverable (at the client's cost) from defaulting package contractors[582].

'Construction management' is a similar fast-track system but with packages awarded by the client direct and with a consultant construction manager in place of a management contractor. The construction manager does not warrant the packages at all but owes a professional duty of care equivalent to that of other consultants[583].

Neither model provides expressly for preconstruction phase joint working by design consultants with the management contractor, construction manager or package contractors. Such joint working could be incorporated by agreement, and is all the more desirable in view of the risks to the client of a limited and fragmented warranty where there is no main contractor assuming liability for the default or insolvency of the specialist contractors and suppliers.

---

[582] JCT Management Contract, clause 3.21.
[583] JCT CM.

### 9.2.4 Single-stage tendering

A large number of clients remain attracted to the single-stage procurement and contracting model whereby design consultants provide detailed designs by reference to which cost consultants prepare detailed pricing documents, on the understanding that to invite tenders against these documents will obtain competitive fixed price bids from prospective main contractors upon which the client can rely. It is attractive for design consultants to enjoy the freedom to develop their ideas to a greater level of detail and it is extremely attractive for clients to have the prospect of cost certainty created by fixed priced bids.

The client's perceived need for an early fixed price, and its reliance on the robustness of such price and the consultant designs on which it is based, can override proper consideration of the potential benefits to be obtained through agreeing and managing main contractor participation in design review and other aspects of project preparation. The continued popularity of single-stage tendering may owe less to its track record in achieving efficiencies and avoiding claims and disputes than to the cautious advice of project managers and cost consultants regarding available alternatives, and to the wish of design consultants to maintain control of the design process for as long as possible.

For the client and consultants to retain exclusive possession of designs, costing and risk assessment until the latest preconstruction stage under single-stage tendering suggests that the client is in control. However, MacNeil noted that to tender this unilateral planning of the project to contractors 'on a take-it-or-leave-it basis' denies the opportunity for mutual planning[584]. He stated that simply agreeing a price without joint examination of all relevant issues constitutes 'allocative planning' which is 'a process heavily laden with conflict'[585]. A major risk inherent in this approach is the opportunism that it encourages among main contractors.

In the view of Bennett & Pearce, single-stage tendering has failed as a procurement system 'because it provided no overall direction, reducing everyone involved to defending their own interests'. They acknowledged the attraction to clients of 'the simplicity of inviting competitive bids', encouraged by 'professionals with a vested interest in old ways of working' but suggested that these clients are 'All too often ... sadly disappointed as they discover that claims, delays, defects and disputes make this an expensive and ineffective approach'[586].

Single-stage tendering does not in fact achieve the price certainty it claims. In a 1975 NEDO survey of predictability between contract

---
[584] MacNeil (1974), 770/771.
[585] MacNeil (1974), 777.
[586] Bennett & Pearce (2006), 7.

prices and final prices, single-stage open competition was found to be the least likely to provide predictable results: only 56% of projects were completed within plus or minus 5% of the contract price[587].

Also, attempts to follow a single-stage tender process with a period of negotiation in order to obtain early contractor input have often failed because they lose the sense of mutual planning and simply result in a hardening of differing positions[588]. Nevertheless, the strong attraction of single-stage tendering is undeniable and remains the biggest barrier to early contractor involvement under preconstruction phase agreements. The risks of single-stage tendering have been highlighted consistently since the Banwell Report in 1964[589]. The persistent misconception that any arrangement other than single-stage tendering should be seen as unusual and high risk was the primary rationale for this book.

## 9.2.5 Project funding

In certain projects, the client's focus will be on obtaining payment of money, for example the private sector funding required for a PFI project or a capital receipt dependent on a related land transaction. This can distract the client's attention from the potential financial and other benefits to be obtained through agreeing and managing preconstruction phase processes.

PFI projects involve extensive preconstruction phase processes which are closely connected to negotiation and financial closure of the PFI deal. None of the preparatory processes for a PFI project are contained in the standard PFI contract form[590]. They are instead undertaken at risk by the service provider and in turn its consultants, main contractor, subcontractors and suppliers.

PFI is structured around private sector organisations funding the project and the client paying a unitary charge following completion of the built facility, a model which assumes that design, construction and ongoing operational risks are passed over to the private sector. PFI documentation reflects this approach and will usually include a building contract that is designed primarily to pass significant design and construction risk down from the private sector provider to its main contractor and that does not provide for any conditional preconstruc-

---

[587] NEDO (1975), Section 5.5, 43. See also Chapter 10, Section 10.4 (Support for two-stage pricing).
[588] See Chapter 2 Section 2.4 (The planning functions of contracts) as to the techniques by which negotiation can be avoided and also Section 2.7 (Risk and fear of opportunism) regarding the risk of contractor opportunism.
[589] See Chapter 4, Section 4.3.4 (Prices and contractor selection).
[590] For example, SoPC4.

tion phase appointment[591]. Nevertheless, thorough preconstruction processes need to be undertaken at an early stage in order to ensure a robust financial model on the basis of which the private sector team can finalise an acceptable unitary charge in advance of financial close. PFI projects often involve complex buildings where joint preconstruction phase working between the design consultants, main contractor and specialist subcontractors and suppliers is essential to arrive at reliable figures, deadlines and risk profiles. The mismatch between this need for joint working and the client's focus on risk transfer means that preconstruction phase activities are likely to take place without the involvement of the client during the periods when the private sector provider and its team are preparing and negotiating their bid. A report by Be noted that 'In the past Government saw risk transfer as a cornerstone of PFI. Thus PFI has focused on risk abdication rather than collaborative risk management'[592]. However, Be also reported that 'The more successful PFI projects have allowed close interaction between building users and a broader range of the supply chain at an early stage in the consideration of designs'[593].

It may be that a financial model demands a fully-packaged design and build price ahead of main contractor engagement, in which case the scope for achieving benefits from a preconstruction phase agreement is limited, whether the driver is private finance under PFI or capital receipts from the sale of land. However, in cases where the same team plan to work together on successive projects, it may be beneficial to integrate successive project financing arrangements or successive land transactions with joint preconstruction phase processes whereby team members can help the client achieve construction cost savings that create an improved financial model for PFI or more profitable land transactions[594].

## 9.3  Procedural obstacles

### 9.3.1  Constitutional or regulatory constraints

If a client has constitutional or regulatory constraints that demand selection of main contractors according to lowest fixed price only, then this will clash directly with early selection of the main contractor by value and will undermine attempts to involve the main contractor in

---

[591] For example, SoPC4.
[592] Be PFI (2003), 11.
[593] Be PFI (2003), 4.
[594] See also Chapter 7, Section 7.5 (Frameworks and the Private Finance Initiative), in particular as regards the Building Schools for the Future programme.

preconstruction phase processes. There are, however, often ways to justify review of such constraints if it can be demonstrated that two-stage procurement is consistent with probity and achievement of best value.

For example, lowest price may be the basis chosen for main contractor selection by those clients who are bound by the Public Contracts Regulations[595]. However, where selection pursuant to the Public Contracts Regulations is on the basis of the 'most economically advantageous tender', there is no inconsistency between these regulations and the early conditional selection of the main contractor to undertake preconstruction phase activities[596]. In addition, there is no requirement under the Public Contracts Regulations that evaluation criteria for the most economically advantageous tender should include a fixed price for the project, provided that the main contractor is being selected to take overall responsibility for that project and has demonstrated objectively against the declared criteria the ways in which it is best placed to deliver that project.

The need for further competitive processes in respect of the selection of subcontractors and suppliers after creation of an early preconstruction phase agreement does not necessarily give rise to those competitive processes themselves being subject to Public Contracts Regulations (2006). If set out in contractual mechanisms which do not affect the overall agreed responsibility of the appointed main contractor, such supplementary competitive processes should be seen (in terms of public procurement law) as equivalent to the well-established systems for firming up provisional sums under construction phase building contracts. The only exception to this rule would arise if the client was seeking to specify or nominate a subcontractor or supplier, in which case the need for a further public procurement process under a preconstruction phase agreement would be exactly the same as the need for an equivalent process in the event of nomination under a construction phase building contract[597].

The Public Contracts Regulations also provide the first statutory recognition of framework agreements and, while limiting them in time (to four years), provide clarity that public sector clients are entitled to establish such framework agreements in relation to the procurement of both works and services[598]. Accordingly, an early conditional contractor appointment leading into a later unconditional construction

---

[595] Public Contracts Regulations (2006) requiring and regulating specific competitive procedures in respect of the award by 'contracting authorities' of contracts of a value over a stated threshold.
[596] Public Contracts Regulations (2006), 30.
[597] See also Chapter 4, Section 4.5.3 (Joint selection of subcontractors and suppliers).
[598] Public Contracts Regulations 2006, Regulation 19.

phase building contract can be compliant with the Public Contracts Regulations whether in respect of a single project or a series of projects governed by a framework agreement. However, if a preconstruction phase appointment is entirely separate from the construction phase building contract, it is arguable that there are two different contracts and consequently that two separate procurement exercises need to be undertaken, one for preconstruction phase services and another for the construction phase works.

It is now permissible under the Public Contracts Regulations also to take into account economically advantageous proposals from tenderers in respect of environmental characteristics where relevant to the subject matter of the contract[599]. However, it is unlikely that such proposals can be fully developed for evaluation as part of a single-stage tender process. For those clients who wish to appoint contractors who are capable of making contributions to the project by way of added value through environmental proposals, an early appointment under a conditional preconstruction phase agreement will provide the opportunity for the cost and practicality of such proposals to be considered by the team as a whole and to be developed into a practical set of propositions before the client is required to give unconditional approval for the construction phase of the project to proceed.

## 9.3.2 Cost and time to create agreements

A disincentive to creation of a preconstruction phase agreement can be the cost and time involved in drafting and agreeing its terms. Cox & Thompson suggested that in relational contracting the use of a contract could have 'the effect of adding unnecessary transaction costs'[600]. A study by Reading University found no evidence of lower procurement costs in creating partnering or framework arrangements, but noted 'the expectation of the parties that this up-front investment results in lower costs downstream'[601]. An obvious means to reduce the time and transaction cost of creating a preconstruction phase agreement is to use a published standard form[602].

When requested to enter into a new type of agreement the team members are likely to seek separate legal advice or at least a view from their professional and indemnity insurers. Advice will vary and some can afford such advice more readily than others. Hence, the process of

---

[599] Public Contracts Regulations 2006, Regulation 30(2).
[600] Cox & Thompson (1998), 83.
[601] Constructing Excellence (2006), 3.
[602] For example, PPC2000 Project Partnering Agreement or a suitable adaptation of the NEC3 Professional Services Contract, as considered in Chapter 6, Sections 6.2.3 (NEC3) and 6.2.4 (PPC2000) and Appendix B.

finalising a preconstruction phase agreement can create imbalances and potential divisions among team members at the very time that they are trying to establish joint ways of working and a common purpose.

The CIC proposed that one answer to this problem would be the appointment of a joint adviser accountable to all team members and recruited 'from any construction profession, or even the legal profession' according to his or her individual skills and expertise[603]. Such an adviser could 'prepare (on an even-handed basis) the documents that record the team's commitments, procedures and expectations'[604].

The Arup report for OGC recognised the independent role of the partnering adviser in shaping the project deliverables and in resolving disputes. It stated that 'The process for the development of the Partnering Documents is supported by the Partnering Adviser who can assist in resolving differences and how the PPC method is to be adopted'[605].

### 9.3.3 Finalising integrated agreements

Another procedure that can be off-putting is the need to reconcile an integrated set of preconstruction phase agreements entered into with the main contractor and various consultants. This requires each team member to conclude and sign an agreement that is consistent with those of other team members, challenging enough when appointing a team of consultants to prepare for single-stage tendering and more challenging when the main contractor is added to the team. The effort involved in achieving reconciliation and integration of the interests of project team members through a full set of consistent agreements should not be underestimated. Milgrom & Roberts noted that 'When there is a diversity of interests, even moderate-sized groups often find it impossible in practice to reach a unanimously acceptable decision'[606].

The use of a multi-party agreement can assist in the early conclusion of mutually acceptable terms among the members of a project team[607]. Where the parties are asked to agree a single set of terms and

---

[603] CIC (2002), 16 Note 13. The CIC call this joint adviser a 'partnering adviser' and the role is expressly provided for in PPC2000, clause 5.6 of the Partnering Terms.
[604] CIC (2002), 16, Note 13. See also Chapter 9, Section 9.6 (The role of the partnering adviser).
[605] Arup (2008), 44, 45.
[606] Milgrom & Roberts (1992), 145. See also Chapter 2, Section 2.3 (Effect of the number of parties).
[607] The PPC2000 Project Partnering Agreement is a single preconstruction phase agreement to be signed by the client, project manager, cost consultant, design consultants and main contractor, and can also be signed by certain subcontractors and suppliers. Multi-party agreements have since appeared in Perform 21 (Perform 21 PSPCP) and JCT CE (JCT CE Project Team Agreement).

conditions and a single integrated programme, they can be more pragmatic in accepting a compromise of their preferred position because it is evident that an equivalent compromise is being accepted by all other team members. By contrast, the negotiation of two-party contracts is often hindered by concerns that one team member's contract is less favourable than other, unseen two-party contracts being finalised with the remaining project team members.

### 9.3.4 Concerns as to conditionality

A major block to the use of preconstruction phase agreements will be the concern that, because they are conditional, they leave scope for opportunism that will leave one party disadvantaged by another's self-interested behaviour. Milgrom & Roberts argued that 'Concern with the possibility of being disadvantaged by self-interested behavior that an incomplete, imperfect contract does not adequately control may prevent agreement being reached in the first place. It may also inefficiently limit the extent of cooperation that can be achieved'[608].

A client obstacle to use of preconstruction phase agreements may be the concern that they are open-ended and that the main contractor may delay start on site irrespective of the client's construction phase deadlines. The corresponding main contractor concern is that the client might extend the preconstruction phase irrespective of limits on the main contractor's planned and priced resources[609]. In either case, these risks of unclear and uncontrolled commitments are far greater if preconstruction activities are undertaken in the absence of a preconstruction phase agreement which states specific activities and deadlines[610].

There is no guarantee under a preconstruction phase agreement that the progressive development of designs, supply chain packages and prices will give rise to a project that is compliant with the client's brief, within the client's budget and capable of being delivered within the client's construction phase deadlines[611]. There is also no guarantee that the process of joint risk management will eliminate or reduce risks and their priced contingencies sufficient to achieve an apportionment of risk acceptable to all parties and the pricing of such risk within the client's budget[612].

---

[608] Milgrom & Roberts (1992), 133. See also Chapter 2, Section 2.7 (Risk and fear of opportunism).
[609] See Chapter 4, Section 4.3.4 (Prices and contractor selection) and footnote quoting NEDO (1975) case study, and see also Chapter 4, Section 4.3.7 (Concerns as to two-stage pricing).
[610] See also Chapter 5, Section 5.4.7 (Programmes as additional contract documents).
[611] See also Chapter 1, Section 1.2 (Early contractor appointments and project pricing).
[612] See also Chapter 1, Section 1.3 (Early contractor appointments and risk transfer).

These uncertainties are compounded by the possibility that new information or changed circumstances during the preconstruction phase may allow one team member to exploit a commercial advantage through opportunism without strictly breaching the terms of the preconstruction phase agreement[613]. Taking into account all of the above, it is important first, that a preconstruction phase agreement remains conditional on clearly agreed terms, allowing for its termination in certain circumstances, until such time as all project team members are satisfied that they have developed and agreed a sufficiently complete contractual arrangement for the construction phase of the project to proceed. To allow a simple option for the client to withdraw at any point and for any reason during the preconstruction phase would demotivate other project team members. However, an unconditional preconstruction phase commitment for the construction phase to go ahead would be the equivalent of a blank cheque and would dissuade clients from making any preconstruction phase commitment at all.

A balance that can be struck is illustrated in PPC2000 where there are stated preconditions for the construction phase to proceed, which include:

- Joint development of designs, supply chain members and prices compliant with the client's brief and sufficient to achieve an agreed price;
- Grant of all required third party consents and compliance with health and safety legislation;
- Fulfilment by the client of related site acquisitions or funding or other previously declared preconditions[614].

Failure to achieve any of these stated preconditions (either completely or to such lesser extent as the team members may agree) allows the client to terminate the appointment of all other team members without proceeding to the construction phase, and to pay only such amounts as have been expressly agreed in respect of activities already performed.

## 9.4 Personal obstacles

Certain obstacles to the early appointment of contractors and the use of preconstruction phase agreements are features of personal attitudes that may or may not be changed through persuasion, education and training. For example, Bennett highlights the mentality of the free

---

[613] See also Chapter 2, Section 2.7 (Risk and fear of opportunism).
[614] PPC2000 Partnering Terms, clause 14.1.

market which can regard even the notion of cooperation or agreement of common interests as being bureaucratic or interventionist and therefore undesirable. He also suggests that in the UK construction industry there are 'deeply rooted structural reasons' that lead it to focus on 'short-termism' and even to be attached to 'adversarial attitudes'[615].

The slow progress in changing attitudes should not defeat innovative approaches to procurement and project management, and it is arguable that such attitudes only strengthen the case for new approaches to be set out in clear agreements. Personal attitudes are likely to be based on personal experience, but as a significant body of positive evidence builds up it becomes very difficult for anyone to deny that alternative approaches demand serious consideration.

Early contractor involvement is the most effective means of breaking down the personal barriers and misconceptions between the client and consultants on the one hand and the contractor on the other hand, who may otherwise draw up battle lines from the moment they first meet, motivated by concerns borne of inadequate information, insufficient preparation and a mismatch of expectations and risk assessments. It is only through joint working during the preconstruction phase that mutual suspicions can be ironed out and practical steps can be taken to agree common methods of working. However, the openness and flexibility required for joint working during the preconstruction phase may threaten the status of parties who hold positions of authority and are accustomed to making unilateral decisions.

Barlow *et al.* noted that problems can arise where individuals need to adopt a consultative approach to projects. They stated, for example, that 'Increased participation can lead to feelings of vulnerability and exposure on the part of managers who were formerly accustomed to leadership. This is partly because participation in close working relationships demystifies the competence of senior personnel'[616].

### 9.4.1 Cynicism

Latham described his 'cynics bestiary' of those 'who do not believe in partnering' as comprising six fundamental types: 'the stick-in-the-mud', 'the jobsworth', 'the one who just doesn't get it', 'the die-hard sceptic', 'the control freak' and 'the young people who don't believe in partnering because they have been fed a poisoned account'[617]. Latham

---

[615] Bennett (2000), 76.
[616] Barlow *et al.* (1997), 16.
[617] Latham (2004).

recognised that new approaches to procurement need to be challenged, but expressed concern that his stated types will remain resistant until offered 'serious training, deep culture change led from the top and continuous reinforcement'[618].

### 9.4.2 Aversion to additional contracts

Where people and organisations seek to avoid contractual commitments, they may justify their actions on the basis of fatalism (e.g. the contract will not really change anything, so it is not worth signing it) or obstinacy (e.g. the team members will do things their own way whatever is written down, so there is no point in a contract). As many practical solutions affecting a project are achieved by pragmatic arrangements apparently outside contract relationships or obligations, there is also often a reluctance to believe that contracts can help to stimulate and support the personal interactions that generate such pragmatism or to clarify the means by which it is put into effect. At best, contracts in this context will be seen as symbolic.

Aversion to contracts may also derive from lack of objective guidance. Often, the client's guide to the choice of a contract form will be its project manager, and that guidance will be informed by the project manager's own training and experience[619]. Consultants who are appointed and paid to coordinate project team members may be reluctant to accept the need for contractual guidance or support and may prefer to rely on their own non-contractual systems and their force of personality to drive the preconstruction phase processes[620]. They may be reluctant to recognise the connections between what they are appointed to achieve and what a preconstruction phase agreement describes[621].

### 9.4.3 Industry conservatism

The construction industry faces continual change and new demands, both in terms of technical advances and new business challenges.

---

[618] Latham (2004).
[619] See also Chapter 8, Section 8.2.2 (The role of the project manager in establishing a procurement strategy).
[620] See also Chapter 9, Section 9.2.4 (Single-stage tendering) and the views of Bennett & Pearce as to the problems caused by 'professionals with a vested interest in old ways of working', Bennett et al. (2006), 7.
[621] See also Chapter 8, Section 8.2.5 (The use of preconstruction phase agreements by project managers).

## Obstacles to Early Contractor Appointments

While it may accept new complex contract structures when demanded by particular business models (such as PFI), the industry will be less enthusiastic if the new contractual arrangements are apparently optional. Conservatism may be borne of continued exposure to changes not all of which are followed through, so that properly documented two-stage procurement (despite its benefits) may be seen as another passing trend. Consequently, if consultants and main contractors consider that clearly agreed processes set out in preconstruction phase agreements are not strictly necessary for the achievement of their business objectives, they will defer or avoid signing up to them if they can.

Conservatism may also be borne out of lack of confidence, leading parties to stick with the most widely used approaches to procurement, for example relying on contract forms that have been repeatedly reviewed in court proceedings even if they are not relevant to the needs of the team and its chosen procurement strategy. For example, Latham observed: 'No doubt some academic lawyers will say of PPC2000, as they also said of NEC3, that it should not be used because it has not been tested in the courts. The absence of courts is a plus, in my view. If you want a document that is regularly tested in the courts, you can use JCT80'[622]. Encouragement to stick with familiar forms of contract, whatever their shortcomings in dispute avoidance, also suggests an 'actuarial' view as to the inevitability of disputes, whereby the longer the users of a new form of contract (such as PPC2000 or NEC3) manage to resolve their differences without a dispute, the greater the likelihood that such a dispute is imminent. Such a view of contracts would lead clients and their teams to resist all proposed improvements.

On the other hand, conservatism can be triggered by excessive confidence, for example where a client has sufficient power in the marketplace to take the view that consultants and contractors will always respond to its needs and offer their the best deals without the client bothering to invest in early planning or team integration, and without the use of joint processes such as value engineering or risk management. To quote the chief quantity surveyor of a leading construction company:

> 'As to where [preconstruction phase agreements] are inappropriate – single-stage tenders and/or where the criterion for winning the work is lowest cost. In the latter case, we would not want to use a pre-commencement agreement because it removes risk, and therefore the opportunity to exploit changes and delays to improve the commercial return'[623].

---

[622] Latham (2002).
[623] Project case study 8, Appendix A.

*Early Contractor Involvement in Building Procurement*

That is fair warning to clients who believe they can seek a single-stage lowest bid price and not expect their appointed main contractor to exploit all available loopholes. Such clients do not know or care what further savings and efficiencies they may be missing[624].

### 9.4.4 Exploiting lack of clarity

Both clients and main contractors may wish to obtain the benefit of early main contractor input to the project, but each may believe that it has a greater commercial advantage in the absence of a preconstruction phase agreement.

The client may believe that it can obtain pre-contract added value without paying for it – although this may not produce best efforts from the main contractor and specialist contractors. In fact, the absence of contractual terms may have the opposite effect, removing any limit on the client's liability for abortive costs incurred during the preconstruction phase, particularly if representations have been made to the main contractor by the client as to the construction phase going ahead[625]. By contrast, a preconstruction phase contractual commitment can reserve to the client the right to terminate the preconstruction phase agreement if any agreed preconditions are not met for the construction phase to proceed, and can specifically exclude consequent claims by any party for loss of profit and other amounts beyond those specifically agreed to be paid[626].

From the main contractor's point of view, the absence of a preconstruction phase agreement to govern early dealings with the client may allow more room for manoeuvre in negotiating higher main contract prices and lower subcontract prices. Such a main contractor may also think that it will be able to secure other commercial benefits such as additional time or lower performance requirements, particularly in an atmosphere of brinkmanship as it becomes time-critical for work to start on site. Whether such an attitude achieves a successful project or a good reputation is doubtful.

In all these cases, the client or the main contractor may be willing to gamble a lack of contractual clarity and security against the possibility of exploiting changing circumstances. Such a decision requires a commercial judgement to be weighed against the benefits of a properly structured approach.

---

[624] See also Chapter 3, Section 3.9 (New procurement procedures or gambling on incomplete information).
[625] See *Baird Textiles Holdings Limited* v. *Marks & Spencer* Plc (2001) EWCA Civ 274, regarding the unpredictable impact of an unwritten agreement.
[626] See for example PPC2000 Partnering Terms, clause 26.1.

## 9.4.5 Perceived bureaucracy

Cox & Thompson observed that one problem when introducing new forms of contract to the marketplace, where they govern project management processes, is the perception that they might 'restrict efficient and/or innovative working' and that they might give rise to added overhead costs if they create guidelines which have to be followed in a narrow bureaucratic manner[627]. The alternative would appear to be a free for all which leaves project management processes unwritten and unreliable. It is important not to suffocate innovation, particularly in design[628], but equally it is arguable that efficiency is increased and overhead costs are saved (rather than the reverse) by a contract that establishes agreed roles, commitments, communication systems and deadlines in relation to critical project activities.

Because of the legal formality associated with the building contract, project teams may consider it is less bureaucratic to describe project processes in other documents. These documents may secure the required preconstruction phase commitments, but only if they have contractual status and fit correctly with the remainder of the contract. Otherwise such documents may not be sufficiently clear and comprehensive to be legally binding and to avoid conflicts or ambiguities when read alongside the contract conditions[629].

## 9.4.6 Reliance on personal relationships

Whatever the contractual structure, a project is unlikely to succeed without good personal relationships between team members[630]. An individual can adopt an aggressive, defensive or unhelpful attitude whatever the form of contract, and this can seriously damage the progress and completion of the project. Hence, it is not surprising that construction professionals working for any client, consultant or main contractor will conclude that it is personal chemistry between individuals that is the first priority for building up trust and sensible working relations, over and above the contractual links between organisations.

A good preconstruction phase agreement will deal specifically with the importance of personal relationships, and will assist this by

---

[627] Cox & Thompson (1998), 46.
[628] See also Chapter 5, Section 5.4.3 (Programming consultant design outputs).
[629] Other contract documents are often overridden by the building contract conditions in the event of a conflict (e.g. PPC2000 Partnering Terms, clause 3.6).
[630] See also Chapter 5, Section 5.3.1 (Communication between organisations and individuals).

identifying key individuals and ensuring that they have to work with each other openly and intelligently[631].

However, in a set of preconstruction phase relationships the parties will be aware of underperformance or non-cooperation of any one or more of their number, and this phenomenon can have a significant negative effect. Milgrom & Roberts described this as the 'free rider problem'[632]. They suggested that any team member might calculate that its own misrepresentation or refusal to cooperate on any aspect of the project would be a way of keeping its own costs at a minimum. The effect of this holding back can be insidious in undermining the success of joint working, particularly among a larger number of project team members. The problem of the 'free rider' can only be addressed if that party can be identified and rehabilitated or replaced. The rights and procedures required to achieve this need to be agreed in advance in a preconstruction phase agreement.

In addition, the creation of early contractual relationships can offer a way to test objectively the personal commitment of individuals representing the different parties. It is reasonable to require that if the parties think as a team, they should be willing to contract as a team. If this is tested at an early stage, it increases the ability of the parties to recognise individuals who are not committed and to provide for them to be replaced or supplemented by additional individuals while the project is still in its preparatory phase.

### 9.4.7 Concerns as to partnering

Problems in the construction sector's understanding of partnering, and the means by which to put it into practice, can also cause concern. For example, any party may consider an arm's length contract more enforceable than a partnering contract and may consider that proposals for partnering or collaborative working could blur the boundaries between their role and responsibilities and those of other team members, giving rise to some kind of contractual soup that cannot be properly analysed and may cause additional potential liabilities. This reflects a perception noted by Cox & Townsend that 'human beings seek to avoid responsibility and must be coerced to perform using a strict contract applied in an arms-length manner', an argument they used to explain the rejection by some parties of any form of collaboration as 'not only impossible to achieve but an inappropriate way of doing business'[633]. The answer to these doubts lies in the establishment

---

[631] See Chapter 5, Section 5.3.1 (Communications between organisations and individuals).
[632] Milgrom & Roberts (1992), 161.
[633] Cox & Townsend (1998), 41.

*Obstacles to Early Contractor Appointments*

of a clear and unambiguous partnering contract that spells out the roles and responsibilities of all parties[634].

Another reason for reluctance to establish and implement project partnering arrangements through construction contracts has been a concern that partnering contracts offer a 'softer' warranty from project team members that could, for example, deter private funders. This concern is not borne out in the robust warranty provisions of the standard forms that promote partnering[635]. As to the perceptions of funders, it is worth noting that the first client organisation to sign PPC2000 in its published form was embarking on a £240m programme that was dependent on the support of private finance from Nationwide Building Society[636]. Gabrielle Berring of Paribas stated in relation to partnering:

> 'The downside is no different from the traditional procurement route. Performance risk on both the client and contractor is no different. Provided that a formal partnering agreement is entered into, which provides specific performance targets and legislates for the circumstances where targets are not met, a funder should embrace the partnering approach'[637].

Where there is a concern as to the implications of partnering, a clear contractual approach is essential in order to ascertain who will be involved in collaboration and the extent of that collaboration. Hence, a conditional preconstruction phase agreement can act as a control mechanism for those who are concerned that partnering may go further than is appropriate for the team or the project in question.

## 9.5 Education and training

This chapter has identified a number of project types and circumstances where the traditional single-stage approach to selection and appointment of contractors is likely to be more appropriate than two-stage procurement facilitating early contractor appointment. However, it also questions whether other causes of concern or reluctance regarding a two-stage approach are logical or justifiable.

Certain of the concerns outlined in this chapter can be addressed through education and training, although it is also relevant to consider how much clients, consultants and contractors actually want to know about the types of contract they may be invited to enter into.

---

[634] See also Chapter 8, Section 8.3.5 (Partnering and building contracts).
[635] See for example PPC2000 Partnering Terms clause 22.1 and options in the Project Partnering Agreement.
[636] Project case study 9, Appendix A.
[637] Berring (1999), 44.

R.J. Smith identified the need for 'owner education' to address various attitudes, including arrogance and ignorance[638]. He suggested that clients would change those attitudes if they could be made more familiar with the benefit to cost ratio of adopting modern contracting practices compared to traditional models. Bennett highlighted the corresponding lack of knowledge in the construction industry regarding the available techniques for achieving improved project processes and noted that:

> 'The early use of partnering made it apparent that most managers in the UK construction industry did not have a working knowledge of the techniques needed to analyse their processes, were reluctant to exchange information in a manner that would support benchmarking, had no objective measures of their own performance and only occasionally used creative techniques such as value management'[639].

Latham underlined the importance of education and training to change deeply embedded attitudes and to avoid young people being fed 'a poisoned account of the way the industry should operate'[640].

The role of the contract as a practical guide through project procedures is a new concept for many in the construction industry. For those who instinctively perceive the contract primarily as a means to pursue or defend claims, the establishment of a process contract as a project management tool is likely to be counterintuitive. This highlights the need for training that is practical and project focused in order to illustrate the benefits of the two-stage approach.

The 2007 Nichols report to the Highways Agency expressed concern that, although early contractor involvement (ECI) had been central to the Highways Agency procurement of major projects for several years, nevertheless 'Since ECI was introduced, there has been very little training provided resulting in a lack of commitment from HA staff at all levels. This has resulted in HA lacking the ability to set sensible budgets, challenge Target Prices and manage the process effectively'[641]. Accordingly, the Nichols report recommended urgent action to strengthen the ability of the Highways Agency to manage the ECI process by means of 'recruiting more staff with appropriate contract and commercial management skills and experience; as well as training and developing existing staff in ECI contracting'[642].

---

[638] Smith R.J. (1995), 69.
[639] Bennett (2000), 79.
[640] Latham (2004).
[641] Nichols (2007), 330.
[642] Nichols (2007), 33.

## 9.6 The role of the partnering adviser

Clients, consultants and contractors need to determine whether or not a preconstruction phase agreement serves their commercial interests in the context of a specific project. No organisation will enter into a contract if it has the impression that to do so would be contrary to such commercial interests[643].

Assuming that in principle the parties are willing to enter into a conditional preconstruction phase agreement, there remains the question of how they may need to be guided as to the impact of new contractual relationships and processes, and who they will turn to in the event of any uncertainty, misunderstanding or potential dispute. Each party will have access to its own advisers, but concern as to increasing transaction costs will often dissuade project team members from each seeking independent advice[644].

It is tempting to suggest that if processes of change need additional advice, then they are unlikely to be sustainable. However, the Construction Industry Council Partnering Task Force, when considering how to disseminate partnering knowledge and best practice, concluded that it is unrealistic to assume that project teams can adopt new approaches to procurement, project management and partnering without such advice[645]. Their view was that, while the ideal number of advisers to support new project processes is zero, the next best number is one.

The solution proposed by the Construction Industry Council was the appointment of a single 'partnering adviser' to assist the team as a whole, not only in creating a conditional preconstruction phase agreement but also in implementing its processes, systems and programmes[646]. This idea was carried through into PPC2000, where the partnering adviser is tasked with the 'provision of fair and constructive advice as to the partnering process, the development of the partnering relationships and the operation of the Partnering Contract'[647]. The PPC2000 partnering adviser is also expected to assist 'in the solving of problems and the avoidance or resolution of disputes'[648].

The value of a partnering adviser will depend on his or her acceptability to all team members as a source of independent advice and support, and this in turn will be linked to the experience that the partnering adviser can bring to the project from other projects in order to

---

[643] See Chapter 8, Section 8.3.3 (Commercial features of partnering).
[644] See also, earlier in this chapter, Section 9.3.2 (Cost and time to create agreements).
[645] CIC (2002).
[646] CIC (2002).
[647] PPC2000, clause 5.6(iv).
[648] PPC2000, clause 5.6(vi). See also Chapter 10, Section 10.11 (Preconstruction phase agreements in an economic downturn).

illustrate how other teams have dealt with similar issues. The Association of Consultant Architects has established an independent Association of Partnering Advisers with accreditation based on relevant experience, and a copy of their Code of Conduct is set out in Appendix F.

In most aspects of procurement and project management, clients look to their project managers for professional guidance to take them through new processes and procedures. There are many project managers who are also experienced in partnering. However, it is arguable that clients, consultants and contractors may also wish to look beyond the project manager for a source of additional independent advice and support. In these circumstances, a specialist partnering adviser can advise on the appropriateness of partnering to a particular project, can help to ensure that the potential benefits of early contractor appointments are fully realised and can guide the team so that the obstacles described in this chapter do not get in the way.

# CHAPTER TEN
# GOVERNMENT AND INDUSTRY VIEWS AND EXPERIENCE

## *10.1  Introduction*

Whatever the reasons for resistance to two-stage procurement and the use of preconstruction phase agreements, it is likely that the status quo of single-stage tendering will remain the industry norm. If so, the question to be addressed is whether the drive for alternative approaches to improve project procurement that emerged following *Constructing the Team*[649] and *Rethinking Construction*[650] will gradually dwindle and die. This chapter will examine the sources of support in the UK Government and construction industry upon which the future of early contractor appointments under conditional preconstruction phase agreements is likely to depend.

## *10.2  Importance of government and construction industry support*

It is not straightforward to change familiar methods of construction procurement and contracting even where there is a compelling case for exploring alternatives. Sustained and influential support is required for any such change to be accepted and implemented. A powerful group of clients, namely central and local government, have encouraged and implemented early contractor appointments and have increasingly recognised that preconstruction phase processes and new relationships (including partnering and frameworks) are necessary to get the best results from the construction industry. Many private sector clients have adopted an equivalent approach, although it is harder to obtain a cohesive picture of private sector trends. Demonstration projects collated in the Egan Report in 1998[651] and by Constructing Excellence in

---

[649] Latham (1994).
[650] Egan (1998).
[651] Egan (1998).

2004[652] reported that private sector clients such as Argent, Whitbread and BAA are increasingly utilising early contractor involvement.

Main contractors are flexible about methods of procurement, and to some extent about contract forms, provided that they can make a profit. They have made considerable efforts to attract early appointments, whether for single projects or (preferably) under frameworks, as these help to stabilise their order books and supply chain arrangements. For so long as their preconstruction phase efforts improve their business, main contractors are likely to help influence project teams to move away from fixed price single-stage tendering[653].

## 10.3 Support for contractor and subcontractor design contributions

A 1975 NEDO report stated that 'Contractors can make a contribution to the design of projects, other than via design and construct projects, in the right circumstances and given the time and inducement to do so'[654]. The NEDO report identified the following circumstances where in its view contractor design contributions would be most effective:

> 'Projects where some form of proprietary building system is being used; serial contracts where the contractor can provide feedback to the design team from experience on an earlier contract; where the method of construction to be used is new or complex; where the construction methods and plant employed are central to finding the most economical design solutions; and where time is of such priority that the construction programme must be compressed'.

This description covers a lot of projects[655].

Early contractor design contributions also reflect aspirations of certain design consultants such as Mott MacDonald, whose director John Hayward was quoted by CIRIA in 1998 as stating 'Our preferred approach, particularly on more complex projects, is to select the contractor who will construct the project at concept stage, that is before serious development of the design. The primary objective is optimise "buildability" and thus improve delivery against time and cost criteria'[656].

---

[652] CE (2004).
[653] Several contractors were members of the task force that produced CIC (2002) and numerous contractors are members of Constructing Excellence and the Housing Forum. See Constructing Excellence website.
[654] NEDO (1975), Section 7.24 and 72.
[655] NEDO (1975), Section 7.24, 72 and 73.
[656] CIRIA (1998), 23 Section 2.4 (Potential for better design and specification).

The Office of Government Commerce included in its 'Critical Factors for Success' the development of 'An integrated project team consisting of client, designers, constructors and specialist suppliers, with input from facilities managers/operators' and 'Design that takes account of functionality, appropriate build quality and impact on the environment'[657]. This theme emerged strongly in the 2005 National Audit Office report which specifically recommends that clients should 'Secure the early and continued involvement of the main contractors and key specialist suppliers in the design of the project' in order to achieve 'Greater certainty and control over delivery to time, cost and quality … creating an environment for innovation in, for example, waste reduction, and maximising the benefits from value engineering'[658].

The audited results of early contractor and specialist involvement on government projects are impressive, in terms of the financial savings and other benefits gained from contractor involvement in design linked to joint costing and risk management under two-stage pricing. For example, the Environment Agency reported savings of £4.4 m of capital costs (3.1%) on its projects and identified further potential savings of £5.8 m in the first nine months of 2004 through 'innovative value engineering arrived at by integrated teams working together at the early stages of projects to reconsider proposed flood defence schemes'[659].

## 10.4 Support for two-stage pricing

Two-stage pricing was the subject of analysis in the 1975 NEDO report where it was recognised as a means of 'formalising the overlap of design and construction phases … for one-off projects of large-scale or complexity'[660]. As to the concern that two-stage tendering allows a main contractor to change its level of pricing, the NEDO report found that, on the contrary, two-stage tendering produced the most predictable results in terms of contract prices corresponding to final prices 'if the selection process has been properly managed and documented (for example, by an elemental cost plan), if a reliable basis of pricing has been established by the client's cost advisers and if there are no significant changes in the client's brief or design concept'[661]. NEDO conducted a survey of predictability between contract prices and final prices, analysed according to the method of main contractor selection, which concluded that two-stage tendering was the most likely to produce predictable results (82% of projects successful within plus or

---

[657] OGC (2007), Construction Projects Pocketbook.
[658] NAO (2005), 29.
[659] NAO (2005), 71.
[660] NEDO (1975), Section 7.31.
[661] NEDO (1975), Section 5.42, 50.

*Early Contractor Involvement in Building Procurement*

minus 5% of the contract price)[662]. However, NEDO recognised that the two-stage procedure does call for 'a greater input from the client or his advisers than simple competition'[663].

Bennett & Jayes observed that if project team members are paid a fair profit and an appropriate contribution to their overheads and other costs, this will always make it much easier for them to concentrate on acting in the best interests of the project and eliminating unproductive activities[664]. In 2003, the Office of Government Commerce specifically recommended, as a basis to achieve an integrated project team, the use of 'modern commercial arrangements based on target cost or target price with shared pain/gain incentivisation'[665]. The National Audit Office also endorsed an approach by which 'The client and supply chain agree a guaranteed maximum price, working to agreed margins with full open-book accounting procedures in place' which it perceives 'builds trust, helps to overcome the adversarial approach to construction and leads to rapid conflict resolution'[666]. It offered the example of the Milton Keynes Treatment Centre, where for three months the main contractor worked on a fee basis developing options for the hospital to consider, and then offered a guaranteed maximum price for delivering the approved design, as a result of which a £15m budget was reduced to a cost of £12m without comprising user requirements or causing delays[667].

However, the establishment of greater confidence in a two-stage pricing approach is dependent on detailed explanations for the benefit of clients and their cost consultants of exactly how such pricing works in practice[668].

## 10.5 Support for selecting contractors by value

The full range of activities to which main contractors can add value were identified by the CIRIA in 1998 as being:

- 'In better teamwork – critical to the success of any project;
- In better programming – shorter project delivery times, or better fit to client constraints;

---

[662] NEDO (1975), Section 5.5, 43.
[663] NEDO (1975), Section 5.42, 50.
[664] Bennett *et al.* (1998), 59.
[665] OGC (2007), Procurement Guide, 05, 5.
[666] NAO (2005), 61, Case Example 7 (NHS Estates and Procure 21 – facilitative support for inexperienced clients).
[667] NAO (2005), Case Studies, 33.
[668] The publication of the PPC Pricing Guide (2008) offers one form of such guidance.

*Government and Industry Views and Experience*

- In better design and specification – better buildability and more effective sourcing;
- In better care of the environment – less waste and damage, better public perceptions;
- In better budgeting – greater sensitivity to the market, specialist knowledge;
- In better management of risk and value – participation in risk and value management'[669].

Among the above opportunities identified for added value arise a number of means to reduce cost, such as contractor contributions to 'shorter project delivery times', 'more cost effective sourcing', 'less waste', 'better management of risk and value', as well as a contractor role in 'better budgeting'. If the main contractor is to be closely involved in trying to save money for the client, this suggests that its quoted price will only be a starting point in the process. This in turn points to the importance of selecting a main contractor on a basis other than a fixed price alone.

CIRIA provided guidance on selecting contractors by value and recognised that this requires 'a greater upfront commitment than straightforward tendering on price', with the need for clients and their advisers:

'To invest the time and money necessary:

- Thoroughly to work through and prioritise what they are seeking to gain from a project;
- To set projects up to enable contractors to contribute the maximum value;
- To identify relevant criteria for their selection;
- To gather information to enable these criteria to be applied'[670].

The sequence of the above guidance is important. In order to run a competitive process for selection of a main contractor other than on the basis of price, the client needs to consider what other criteria are important to it that could be assessed objectively when comparing alternative bids.

The Cabinet Office in 1995 recognised that 'The best projects [we saw] and the best private sector clients put time into getting the right team. They assessed the quality of the individuals, their ability to work together and their experience'[671]. The Cabinet Office were concerned

---

[669] CIRIA (1998), 14.
[670] CIRIA (1998), 8. See also Chapter 4, Section 4.3.5 (Two-stage pricing) setting out the CIRIA qualitative selection criteria.
[671] Efficiency Unit (1995), 253, 76.

that public sector clients frequently put together teams for the wrong reasons with undue emphasis on lowest price or expediency. They recommended, for example, that interviewing the individuals who will actually work on the project should be normal practice[672].

The Office of Government Commerce includes in its 'Critical Factors for Success' 'Award of contract on the basis of best value for money over the whole life of the facility, not just lowest tender price', 'An integrated process in which design, construction, operation and maintenance are considered as a whole' and 'Procurement and contract strategies that ensure the provision of an integrated project team'[673].

The National Audit Office in 2005 recognises that value for money savings can be achieved through the early development of an integrated project team. They state that 'Several of the case studies in this report show how collaboration and integrated team working reduces costs and improves performance. Whether carrying out single projects or large programmes, departments should consider a two-stage procurement technique to bring contractors into the design process at an early stage[674].

## 10.6  Support for joint risk management

In 1995 the Cabinet Office found that 'Many Departments are not investing enough effort to address risk when a project is being conceived' and identified 'two projects which ran into trouble because environmental problems were not identified before construction started'[675]. In their view, risk management comprises identification and analysis of risks as well as 'planning how risks are to be managed through the life of the project to contain them within acceptable limits'[676]. They stated that:

> 'The planning phase should result in production of the Risk Management Plan. This should define acceptable levels of risk in areas of cost, time and quality; detail the risk reduction measures to be taken to contain risks within these levels; outline cost-effective fall-back plans for implementing if and when specific risks materialise; identify the resources to be deployed for managing risk [and] explain the roles and responsibilities of all parties involved in risk management'[677].

---

[672] Efficiency Unit (1995), Section 254, 76.
[673] OGC (2007), Construction Projects Pocketbook, 1.
[674] NAO (2005), 68.
[675] Efficiency Unit (1995), Section 154, 56.
[676] Efficiency Unit (1995), 56.
[677] Efficiency Unit (1995), 56.

*Government and Industry Views and Experience*

A 1991 ICE/DETR report recommended that 'Clients should take an active role to ensure that key elements of management (such as risk management) are put in place early during pre-project planning'[678]. In 2003 the Office of Government Commerce included in its 'Critical Factors for Success' a system of 'Risk and value management that involves the entire project team, actively managed through the project'[679].

## 10.7 Support for greater client involvement

Closer client involvement in projects was envisaged by Banwell *et al.*[680]. A 1975 NEDO report highlighted that: 'A construction project requires collaboration between the participants: the client, the design team and the construction team. Each has an important role to play and the quality of the interaction between them, the degree of "teamwork", will be one of the deciding factors in ensuring the successful outcome of the project'[681]. The NEDO report went on to identify in particular 'the client's role in establishing his own objectives; and in establishing a proper brief for the project and clear reporting arrangements and lines of communication to enable him adequately to monitor the overall progress of the project through the design and construction phases'[682].

As a cautionary note, the NEDO report emphasised that client involvement 'should not be allowed to become client interference with the responsibilities of the other participants'[683]. Client responsibility was also emphasised in 'Construction Procurement by Government' which proposed client involvement through the project owner who should 'provide the necessary leadership and be clearly accountable for delivering the project requirements in accordance with approvals given', including establishing the budget, the organisation structure and communication processes, ensuring involvement of users and stakeholders, ensuring that an appropriate brief is developed, establishing a progress and reporting procedure and also approving changes and dealing with problems and disputes through to post-completion evaluation[684].

In 2003, the Office of Government Commerce included in its 'Critical Factors for Success' 'Leadership and commitment from the project's

---

[678] ICE/DETR (1991), 27.
[679] OGC (2007), *Construction Projects Pocketbook*, 1.
[680] As described in Chapter 5, Section 5.2.1 (The need for client involvement).
[681] NEDO (1975), Section 3.1, 25.
[682] NEDO (1975), 1975, Section 3.1, 25.
[683] NEDO (1975), Section 3.8, 26.
[684] Efficiency Unit (1995), Sections 35, 36, 73, 75, 76 and 83.

Senior Responsible Owner' and 'Involvement of key stakeholders throughout the project'[685]. It specifically identifies the client in its integrated project team. In 2005, the National Audit Office made the clear recommendation that 'Departments need to develop and support well focused and capable public sector construction clients', able to offer 'intelligent central support' and provide 'effective and consistent leadership throughout the course of construction projects'[686].

## 10.8 Government and industry views on partnering

In 1998, Egan identified numerous examples of successful private sector partnering and positively promoted this as an approach to public sector projects[687]. Since then, partnering has increasingly become a significant plank of government construction policy, promoted in the Office of Government Commerce Achieving Excellence in Construction guidance[688], whose recommendations included 'Introduce partnering into all property and construction projects'[689].

The 2003 Local Government National Procurement Strategy, while not making partnering a precondition for funding, clearly encouraged the adoption of partnering and collaboration as a basis for local government strategy in the procurement of capital projects[690]. By 2005, the NAO had also made its position clear, repeatedly underlining its endorsement of partnering.

For example, Latham's foreword to NAO (2005) stated that 'Best practice is about partnering, collaborative working and stripping out of the equation at the earliest possible stage those costs which add no value' and should extend not only to 'first tier contractors' but also to 'specialist contractors' many of whom have significant design responsibilities[691]. The National Audit Office described partnering as 'A structured management approach designed to promote collaborative working between contracting parties', whether applied to a single project or a series of projects, with the aim of achieving continuous improvement from one project to the next[692]. It went on to suggest that organisations adopting a partnering approach should:

---

[685] OGC (2007), Construction Projects Pocketbook.
[686] NAO (2005), Section 3, 9.
[687] Egan (1998), 20.
[688] OGC (2007), Procurement Guide 05.
[689] OGC (2007), Procurement Guide 01, 16.
[690] ODPM/LGA (2003), Partnering and Collaboration, 25–33, and Guidance on Partnering, 56–60.
[691] NAO (2005), 1.
[692] NAO (2005), 6.

- 'Work in a positive no-blame whole team environment;
- Provide early warning to each other of any matters than could affect the achievement of the project objectives;
- Use common information systems and work on an open book basis including showing the elements of contingency and risk allowances added to costs, prices and timing of all future work;
- Have incentives for delivery based around pain/gain share arrangements'[693].

The first two recommendations are features of a well-structured communications strategy and the latter two are features of a well-structured pricing model. All four are arguably dependent on a clear agreement between the client and the consultants and the main contractor during the preconstruction and construction phases of a project. It is significant that the RIBA Outline Plan of Work published in 2007 recognised a different sequence of work stages for use by architects engaged on partnering projects, whereby the main contractor appointment is tendered prior to concept design and specialist appointments are tendered prior to technical design[694].

The 2007 Ministry of Defence Partnering Handbook stated that 'Partnering gives us a better chance to deliver on target against time, performance and cost by working together with our suppliers to solve problems, manage risk and deliver successfully'[695]. The National Audit Office review of the Job Centre Plus Office programme commented that:

'The partnership approach to the majority of the supply chain was one of the primary changes to the project's procurement arrangements. Stakeholders have commented that the project was successful in creating a completely open and non-confrontational environment that allowed for Regional Works Contractors, subcontractors and suppliers to work with the Department as a single integrated team'[696].

## 10.9 Government views on preconstruction phase agreements

The use of the conditional preconstruction phase agreement contained in PPC2000 appears to be increasing. The RICS survey of building contracts in use in 2001 found that PPC2000 accounted for only 1.2%

---

[693] NAO (2005), 5.
[694] RIBA (2007), 2.
[695] MoD (2007), 2.
[696] NAO (2008), 23.

of projects by value[697], whereas the equivalent survey in 2004 found that the PPC2000 share had increased to 6%[698].

Meanwhile, in 2003 the Office of Government Commerce set out the features it expected to see in a government construction contract, including early contractor appointment prior to finalisation of designs and prices[699]. In 2005, the National Audit Office recommended the use of 'forms of contract that embed the principles of collaborative working and good project management'[700]. Their perception of collaborative working was that it should extend to 'the client and the entire integrated team' and that procurement and contracting strategies should embody 'A well-developed capability to identify and manage the construction project risks'[701].

The National Audit Office went so far as to recognise that, as a basic requirement to achieve successful procurement, public sector clients should recognise that 'the form of contract itself is an incentivising force'[702]. This contrasts with a government report ten years previously by the Cabinet Office which stated: 'Many believe that the particular form of contract is irrelevant if the team appointed is determined to find solutions not problems'[703].

The Office of Government Commerce commissioned an independent report from Arup Project Management to compare the respective merits of NEC3, PPC2000 and JCT CE in satisfying the principles set out in its Achieving Excellence in Construction initiative[704]. The report concluded that all three forms of contract would enable parties using them correctly to meet the Achieving Excellence in Construction standards. Significantly, the report identified that: 'The PPC2000 documentation represents a complete procurement and delivery system that is distinct from other forms of contract available'[705]. It recognised that: 'The impetus of the PPC form is for "early contractor involvement"' and stated that: 'This should result in the Client procuring his Constructor at a point in the process where his specialist construction and management skills can have a great impact on the project'[706].

It may be suggested that the UK Government will tell everyone else how to implement best practice, but will fail to follow its own advice. It is therefore encouraging to note, for example, the use of preconstruc-

---

[697] RICS (2001), 18, 19.
[698] RICS (2004), 19, 21.
[699] OGC (2007), Guides 1 to 11.
[700] NAO (2005), 14.
[701] NAO (2005), 67.
[702] NAO (2005), 67.
[703] Efficiency Unit (1995), 75.
[704] Arup (2008).
[705] Arup (2008), 41.
[706] Arup (2008), 40.

tion phase agreements by the Highways Agency when implementing its Early Contractor Involvement (ECI) projects[707]. This initiative was launched as part of the Highways Agency Procurement Strategy in 2001 under a form of contract based on NEC2 governing 'early involvement of the supply chain in the planning and design of projects and service requirements' which it was perceived would lead to 'greater innovation, better risk management, forward investment in staff and plant, and affordable, safer solutions'[708].

The Highways Agency Procurement Strategy made a commitment to teamwork, requiring identification of mistakes and joint working to put them right, but recognised that a teamwork approach 'does carry a specific risk of a lack of clarity for liability and decision-making' and that 'it is therefore essential that contractual roles and responsibilities are clear and understood at the outset'[709]. ECI projects therefore utilise a formal preconstruction phase agreement and supporting timetable. For example, the preconstruction phase under the 'programme timetable' for the A14 Haughley New Street to Stowmarket ECI scheme ran from 14 March 2005 to summer 2007 allowing for the duration of a public enquiry[710].

There are numerous other examples of the UK Government implementing and promoting frameworks that include early contractor and specialist appointments under preconstruction phase agreements. These include:

- HM Prison Service (National Offender Management Service (NOMS)) frameworks (from 2002) governing a national programme of up to £3bn of newbuild and refurbished prisons. These frameworks used PPC2000 to establish and implement preconstruction phase appointments for each project governing joint client/consultant/contractor finalisation of designs, supply chains and prices and joint management of risk[711].
- Job Centre Plus frameworks (2003 to 2006) governing a £737m national programme of government office refurbishment. These frameworks also used PPC2000 to establish preconstruction phase processes and achieved overall savings of 24.8% compared to original cost estimates[712].
- Building Schools for the Future programme (from 2005) providing for Local Education Partnerships to put in place Strategic

---

[707] Highways (2005), Section 2.2 Strengthening the supply chain.
[708] Highways (2005), Section 1.1 Progress against the 10 Principles for Best Value (Early creation of delivery team).
[709] Highways (2005), Section 2.4, Collaboration.
[710] Highways (2005).
[711] *Building Magazine*, 12 November 2004, 11 and 30.
[712] Project case study 10, Appendix A.

Partnering Agreements governing preconstruction phase design development and finalisation of supply chains, prices and construction phase programmes on successive projects[713].
- NCA Housing Initiative (from 2005) under which the Department for Communities and Local Government funded the development of template main contractor and subcontractor/supplier framework agreements setting out preconstruction phase design, pricing, risk management and programming activities for use by consortia of social landlords implementing collective large-scale procurement of housing stock refurbishment[714].

Government support for preconstruction phase agreements and framework agreements can only be expected to continue for so long as they achieve results, and in this respect the Government's own statistics are increasingly positive. An earlier government client improvement study in 1999 found that 25% of government projects were delivered to budget and 34% were delivered to programme[715], whereas by 2005 the NAO report found that 55% were delivered to budget and 63% to programme[716], a significant improvement several years after the Government started to implement its own recommendations regarding two-stage procurement.

## 10.10 Industry experience of preconstruction phase agreements

The Strategic Forum reported in 2001 that, from the several hundred public and private sector demonstration projects that adopted the *Rethinking Construction* principles in the preceding four years, significantly better results had been achieved than the industry average against all the targeted improvements in cost, time, defects, accidents, predictability, productivity and profitability. Examples over the four-year average included client construction costs 6% lower than the industry average, project accident rates 61% lower than the industry average and profit 2% higher than the industry average[717].

Willmott Dixon's first partnering project using a preconstruction phase agreement (under PPC2000) was for the construction of Bleak Hill School in St Helens in 2000. As at October 2006 it had completed 33 projects using preconstruction phase agreements under PPC2000,

---

[713] PfS 2008. See also Chapter 7, Section 7.5 (Frameworks and the Private Finance Initiative).
[714] See NCA Website.
[715] AE (1999).
[716] NAO (2005).
[717] Strategic Forum (2001).

none of which involved any disputes and had delivered a total of £450m of projects under partnering arrangements and frameworks[718]. In 2005, Wates Group won 74% of its work through partnering including framework arrangements and it believes that its 'broad partnership experience translates effectively for private sector clients'[719]. Dave Smith, Managing Director of Wates Group, considered a preconstruction phase agreement to be 'absolutely vital, but only if all the constituent parts [contractor design input, supply chain, joint management of risk, agreement of construction phase programme] are owned'[720].

The Constructing Excellence report 'Demonstrating Excellence'[721] included a graph showing that over the period to 2004 a total of 414 public and private sector demonstration projects compared well on average to the remainder of the construction industry as regards client satisfaction (produce and service), defects, safety, cost predictability (design and construction), time predictability (design and construction), productivity, construction time, environmental impact (process and product), staff turnover, employee satisfaction and qualifications and skills[722]. The National Audit Office 2005 Report identified successes through partnering and collaborative working achieved in frameworks organised by Defence Estates, the Environment Agency, the Highways Agency and NHS Estates[723]. It pointed to a significant factor in this success being 'the move towards implementation of the principles of good construction practice set out in Achieving Excellence and Constructing Excellence'[724].

## 10.11 Preconstruction phase agreements in an economic downturn

The use of preconstruction phase agreements to achieve early contractor involvement, both for single projects and as part of multi-project frameworks, has increased during a ten-year period of relative economic prosperity. It is important to consider how this phenomenon would be affected by a less healthy economy, and whether the new contractual processes and the Government and industry views that support a two-stage approach will stand up to the different demands placed upon them.

---

[718] Latham (2006).
[719] See Wates website.
[720] Project case study 9, Appendix A.
[721] CE (2004).
[722] CE (2004), 2.
[723] NAO (2005), 40.
[724] NAO (2005), 39.

A query raised by Bresnen & Marshall is whether partnering is 'contingent upon a number of commercial and other considerations'[725]. In this context, Barlow *et al.* noted that in France in the early to mid-1980s partnering between main contractors and subcontractors was promoted as a means of increasing efficiency, but collapsed in the 1988 recession when preferred subcontractors broke from these partnering relationships. Barlow *et al.* recognised, however, that in France at that time there was no industry-wide body to promote partnering, as a result of which subsequent initiatives were dissipated and lessons were lost[726]. By contrast, in England the Government and the construction industry have supported best practice bodies such as the Construction Industry Council and Constructing Excellence[727].

It has been suggested that, as a result of the economic downturn during 2008, clients were increasingly returning to single-stage tendering[728]. Yet the same commentator recognised on the other hand that:

'A harsh economic climate could encourage adversarial behaviour, and it is important to recognise that clients moved away from single-stage tendering for good reasons. The process can be wasteful of resources, it separates design and construction and, when tendered on incomplete information, provides an illusory promise of competitive pricing and cost certainty'[729].

The argument was put forward that, despite the risks inherent in single-stage tendering, changing market conditions create an opportunity for clients to use single-stage tendering to obtain more competitive prices. I would suggest that, on the contrary, new economic circumstances giving rise to a possible shift in a balance of power between clients and contractors should be used more scientifically than simply to drive down contractor prices in a single-stage bid. Instead, such circumstances create the necessity for a deeper client examination of underlying costs and related value, and this can best be achieved through a two-stage process.

Cost certainty has never been a notable feature of single-stage tendering, as evidenced by the significant number of cost overruns deriving from changes and claims for delay and disruption[730]. A well-documented fixed price commitment can be produced more

---

[725] Bresnen & Marshall (2002), 232.
[726] Barlow *et al.* (1997), 64.
[727] CIC and CE are both supported by industry membership and CE was originally established (as Rethinking Construction) with government financial support.
[728] Rawlinson (2008), 68.
[729] Rawlinson (2008), 68.
[730] See Chapter 3, Section 3.6 (Causes of claims).

convincingly as a result of a two-stage procurement process, in the confidence that the brief and the project risks have been thoroughly reviewed and accepted by all team members, including the contractor who will build the project[731].

It has been suggested that single-stage tendering offers the 'discipline of completing the design before a contractor appointment takes place'[732]. This ignores evidence, from the Banwell Report in 1964[733] through to the National Audit Office in 2005[734], that contractors and subcontractors have more to offer than simply a price for building someone else's designs. Systems that obtain timely contractor input to buildability and affordability provide greater disciplines in procurement, and a proper two-stage process can include these systems[735].

The argument for single-stage tendering is that in tougher times it applies 'commercial pressure to secure cost reductions'[736]. However, meaningful cost savings can be achieved more effectively and transparently through two-stage tendering. Most single-stage bids rely on estimates from subcontractors and suppliers that are invisible to the client and its consultants[737]. There is no opportunity to drill down into what the subcontractors and suppliers have assumed or to discover what they can offer to challenge design or risk assumptions.

However, under a two-stage process, subcontractors bid to a pre-appointed main contractor, with better odds of success and consequently greater motivation to give the bid their best shot. While a main contractor may be tempted to adopt a last-minute negotiating stance designed to improve its commercial return and shift risk to the client, a well structured two-stage appointment can avoid this. For example, the client can agree main contractor profit and overheads from the outset, plus an incentive for the contractor to reduce other costs[738]. This approach, combined with second tier competitive processes for all subcontractors, minimises the need for any negotiation and any related brinkmanship.

Alarmingly, the supposed benefits of single-stage tendering have included the suggestion that 'keeping the client at arm's length over the selection of the contractor's team' should help clarify the allocation of risk[739]. Yet client involvement in finalising contractor and specialist selection is essential to achieve a clear brief and to minimise later

---

[731] See Chapter 4, Section 4.3.5 (Two-stage pricing).
[732] Rawlinson (2008), 68.
[733] Banwell (1964).
[734] NAO (2005).
[735] See Chapter 4, Section 4.2.1 (Contractor design contributions).
[736] Rawlinson (2008), 68.
[737] See also Chapter 4, Section 4.3 (Preconstruction pricing processes).
[738] See Chapter 4, Section 4.3.6 (Treatment of profit and overheads).
[739] Rawlinson (2008), 69.

changes[740]. The inference that this involvement blurs the risk position does not reflect the way that systematic two-stage tendering should be undertaken. Allocation of risk following joint supply chain selection with client involvement should not dilute the contractor's responsibility to deliver the project[741].

Finally, it has been argued that single-stage tendering can improve the 'overall speed of project'[742]. Yet joint programming under two-stage tendering encourages elimination of wasted time by clear agreement in advance of deadlines for the interfaces between all parties' construction phase activities[743]. As to the preconstruction phase, the 2007 Nichols report found that early contractor involvement could reduce project preparation time by 30 to 40%[744].

Clients, consultants and contractors have seen ample evidence of the benefits of securing early contractor involvement by means of two-stage tendering. Using this system to add value and foster teamwork can be seen as progress on the evolutionary scale of procurement and its benefits are available in any economic circumstances, so why should a downturn tempt us to revert to something more primitive? Ten years ago CIRIA concluded that selecting contractors by value under two-stage tendering results in better teamwork, programming, design and specification, care of the environment, budgeting and management of risk and value[745]. All these arguments remain valid.

As regards the risk of clients and contractors reverting to lowest price single-stage tendering in response to an economic downturn, *Constructing the Team* author and Willmott Dixon deputy chairman Sir Michael Latham offered these observations:

> 'Returning to the old ways of adversarial behaviour will lead to more conflict between client and contractor, with variations and claims working up the original tender price, as contractors look to make the money that was not in the original tender. If clients go back to the bad ways, the industry will do the same. Instead of focusing on project outcomes their concentration will be on preparing for costly legal claims in court. By turning away from partnering approaches, clients are in danger of throwing away all the best practice that has given projects better predictability of price, quality and delivery'[746].

---

[740] See Chapter 5, Section 5.2.1 (The need for client involvement).
[741] See Chapter 4, Section 4.5.3 (Joint selection of subcontractors and suppliers).
[742] Rawlinson (2008), 69.
[743] See Chapter 5, Section 5.4.5 (Early agreement of construction phase programmes).
[744] Nichols (2007), 32.
[745] CIRIA (1998), 14.
[746] Email to author, November 2008.

## Government and Industry Views and Experience

There is the risk that funding problems and related insolvencies will fuel additional claims and disputes during an economic downturn, and the test of two-stage contracts using a partnering approach will depend on whether they help the team to navigate through these claims and problems in a different way.

The structure of a two-stage contract provides additional agreed information as to standards and costs, clearly agreed dates for all parties' deliverables and closer working relationships. These in turn provide the parties with more information with which to assess the validity of claims, plus someone to speak to in order to seek to resolve such claims within properly structured terms of reference. All this should increase the possibility of a negotiated solution and should reduce the likelihood of a dispute being referred to an external third party.

The case studies set out in Appendix A include seven illustrations of projects that were completed within budget, within programme and according to their specifications without any claims or disputes. However, the more powerful illustration is Project case study 8, where claims and disputes did arise, but where they were resolved between the parties themselves using the assistance of a partnering adviser[747], but without exercising their rights to go to adjudication or litigation.

This chapter has illustrated that the UK Government and the construction industry are increasingly supportive of early contractor appointments under preconstruction phase agreements, on the basis of the demonstrable results achieved by projects procured in this way. It has also offered evidence of continued contractor willingness to enter into such agreements, particular where they are linked to the award of further work. It is therefore reasonable to expect that continued good results will attract more clients, contractors and consultants to explore the potential benefits that such agreements can bring to their projects.

---

[747] See also Chapter 9, Section 9.6 (The role of the partnering adviser).

# CHAPTER ELEVEN
# CONCLUSIONS – THE GOLDEN AGE OF ROCK 'N' ROLL?

Should early contractor involvement and its influence on partnering be seen as a sideshow or as a significant step forward in mainstream project procurement? Will the Latham/Egan years and the new contracts that they have spawned be recalled as the construction industry equivalent of (to quote a Mott the Hoople song) 'The golden age of rock 'n' roll'?

While rock 'n' roll can be traced back to Elvis Presley in 1954, early contractor involvement has its origins ten years later with the less flamboyant figure of George H. Banwell. However, where rock 'n' roll took off and never looked back, early contractor involvement promptly went into a 30-year hibernation, and only through the efforts of Sir Michael Latham in 1994 did we start to wake up to its full potential.

Since then, contracts offering early contractor involvement have come and gone with varying impact. GC/Works/1 Two Stage Design and Build and Perform 21 seem to have vanished from view, whereas PPC2000 has acquired a loyal following and increasing recognition. NEC3 has become a significant force that in turn seems to have prompted JCT to produce an increasingly wide variety of contracts, although neither NEC nor JCT has yet been persuaded to create an integrated two-stage contract form that describes the processes governing conditional early contractor involvement.

Over the years since the Egan Report, sustained government and industry support for partnering and improved procurement has produced a wealth of data to illustrate the benefits of early contractor involvement in practice and to underline the need for its systems to be clearly documented and properly applied. All of this should place early contractor involvement firmly on the agenda as an option to be considered by clients and consultants when formulating a project procurement strategy.

As the construction industry and its clients face the demands of another economic cycle, it should also be remembered that the 1994 Latham Report was inspired by the need for the construction industry to offer better value at a time when the UK was looking for ways to climb out of recession in the early 1990s. A history of fragmented teams, expensive disputes and dissatisfied clients, as revealed by Latham, described an era that none of us would recall as a golden age by any measure.

This book has considered the use of the preconstruction phase agreement as a 'process contract' to describe the means by which the project team can gain from early contractor involvement. In this chapter, I will review the conclusions that can be drawn from the arguments, counter-arguments and case studies that I have reviewed.

## 11.1 Functions of building contracts and their potential to govern project processes

It has been established that a conditional preconstruction phase agreement is nevertheless an enforceable contract, provided that it contains the machinery for moving from incomplete to complete information and is not simply an 'agreement to agree'. It should be noted, however, that such machinery may not deal with every eventuality and that subsidiary matters can be resolved through reasonable behaviour and negotiation. In these matters, the preconstruction phase agreement is in part a relational contract dependent on a project management culture such as partnering that encourages reasonable behaviour.

After consideration of the features of contract law that are relevant to a preconstruction phase agreement, the following observations emerge:

- Preconstruction phase agreements require the parties to recognise gaps in their planning and set out the processes and techniques by which to fill such gaps.
- Such processes need to include methods for dealing with matters outside the parties' control and to set limits as to the flexibility allowed in adjusting to such matters. The preconstruction phase agreement needs to provide that in circumstances beyond the limit of such flexibility the project will be abandoned.
- The larger the number of parties to the preconstruction processes, the more complex are the coordination and planning of structures and processes under a preconstruction phase agreement.
- Some planning functions of a preconstruction phase agreement are allocative (e.g. agreed design deliverables), whereas others are more tentative (e.g. grant of design approvals) as their outcome cannot be predicted with complete certainty. Such planning is an incremental process requiring a series of choices to accommodate and utilise additional information.

- The preconstruction phase agreement is less efficient to the extent that earlier in the project process there are more unknown items, but this needs to be balanced against the greater inefficiency of postponing contractor involvement until it is too late to take corrective action to deal with information that is complete but inaccurate.
- When contractors are required to bid for later appointments, and are therefore dependent in relation to significant issues on negotiation rather than joint working, they are more likely to be opportunistic if placed at a disadvantage.
- Clear preconstruction phase agreements can harmonise divergent commercial interests, even where early activities are not fully rewarded, if they establish rational coordination to achieve common goals.
- Investment in preconstruction phase activities is motivated by a variety of factors and may be increased by the prospect of additional projects.
- Preconstruction phase agreements can offer the means to avoid misunderstandings and to compromise strict entitlements so as to achieve cooperative adaptation in the collective interests of the project team.

In the light of these observations, it is arguable that a preconstruction phase agreement can act as a handbook for performance, managing the early project processes and promoting good practice. As these processes and practices are not so well understood and accepted as to be adopted automatically, such a contractual handbook therefore has an important role to play.

## 11.2 The assumption in standard form building contracts of complete project information and consequent problems and disputes

A review of standard form building contracts governing only the construction phase of projects reveals their apparent lack of conditionality and their assumption of complete project information at the time they are entered into. However, typical contractual provisions dealing with provisional sums and responses to change and risk events illustrate that construction phase building contracts are not entirely unconditional or complete.

It may be that the evolution of certain standard forms has been limited by the need for consensus between representatives of different industry interest groups involved in their drafting. However, the case for further development of such contracts, particularly so as to integrate design and construction activities, is underlined by expressions of client dissatisfaction.

Previous studies have demonstrated clearly how the primary causes of claims and disputes can be traced to inadequate preparation during the preconstruction phase, and that construction phase building contracts can do little to protect the parties' interests as they do not exist at the time when they could regulate such preparation.

Based on this evidence, it is proposed that reliance on a procurement model that offers only construction phase building contracts following a single-stage tender creates significant risk of inefficiencies, claims and disputes, and that an alternative approach is necessary.

## 11.3 Preconstruction phase processes that can be improved by early contractor appointments

It is proposed that a suitable alternative procurement model is two-stage procurement with early contractor appointment under a conditional preconstruction phase agreement.

Empirical evidence taken from eight project case studies has been used to illustrate the benefits that can be obtained through the following preconstruction phase contractor contributions:

- Contractor design contributions – the client is not taking full advantage of the contractor's capability if it does not obtain its contributions to verifying affordability and buildability and to developing specialist designs. Closer integration is needed between consultant design processes and those undertaken by the main contractor and by its subcontractors and suppliers. Project case studies illustrate the value of contractor design contributions and commentators suggest that in order to maximise such contributions the contractor should be rewarded for its work.
- Contractor contributions to pricing – notwithstanding the client's eagerness for a fixed price and the need for accurate information to achieve a fixed price, experience points to the difficulty of achieving accurate information under single-stage tendering and leads to the proposed alternative of two-stage pricing. This involves as the first stage early selection of the main contractor on the basis of evaluating limited pricing information and other qualitative criteria, followed by the second stage during which the client and main contractor jointly select subcontractors and suppliers and build up the remaining pricing information. Project case studies illustrate the benefits of two-stage pricing in motivating value engineering,

minimising risk contingencies and assisting construction phase cost management.
- Contractor contributions to risk management – there are unavoidable shortcomings in risk assessments undertaken only by the client and its consultants, and also limited time and opportunity for bidding contractors in a single-stage tender to undertake their own risk assessments. This approach severely restricts any scope for joint risk management, leaving risks priced inaccurately and no time for actions to address those risks prior to start on site. Project case studies illustrate the contributions that contractors can make to eliminating or reducing risks if they are appointed early in the project process. As regards the option of risk sharing during the construction phase, this may be appropriate in certain cases but this is not a substitute for pre-emptive joint risk management.
- Early subcontractor and supplier appointments – it is submitted that early contractor appointments should not stop with the main contractor, but should seek to capture contributions from key subcontractors and suppliers. Project case studies illustrate the contributions that subcontractors and suppliers can make to design and risk management. Where main contractors are not willing to extend their own early appointments to their subcontractors and suppliers, this underlines the need for clearly agreed client-led preconstruction phase joint selection processes and deadlines.

## 11.4 The role of the client, communication and programming of preconstruction phase processes

Preconstruction phase systems that will influence the efficiency of early processes include client involvement, communication systems and pre-agreed programmes. Eight project case studies have been used to illustrate the operation and importance of each.

The central contractual position of the client leads to increased risks if it has insufficient direct involvement in the project to fulfil its role efficiently. Experience shows the particular importance of a clear client role during the preconstruction phase in linking the contributions of consultants and contractors and in clarifying its own requirements and expectations. Project case studies have illustrated the ways that early client involvement, or lack of it, can influence project outcomes.

Most standard form building contracts contain limited communication systems and commentators suggest the need for a more detailed strategy to establish and sustain efficient communication between team members. The success of a project requires communication between individuals as well as organisations, including through properly organised meetings, and it is suggested that this can be achieved through the use of a core group of key individuals with agreed terms of reference providing a forum to explore new ideas and to resolve problems. It is also proposed that an agreed early warning system can encourage team members to notify problems in their own or another party's performance. Project case studies have illustrated the problems and disputes that can be solved or averted through establishment of a core group and early warning system early in the project process.

In recognising the importance of programmes as project planning tools during the preconstruction phase, it is argued that there is a need to agree contractual deadlines for consultant design outputs and other preparatory activities, including the early agreement of a construction phase programme. Such deadlines, where they can be expressed as key dates for deliverables and interfaces between team members, should be included in a preconstruction phase agreement. Project case studies have illustrated the benefits that can be obtained through contractual clarity as to mutually agreed deadlines.

Deadlines for implementing preconstruction phase activities also offer a means to agree in advance a contractually binding construction phase programme stating realistic key dates for those construction phase activities where team members are reliant on each other.

## 11.5 Contractual and non-contractual options to govern preconstruction phase processes

Having demonstrated the potential benefits of early contractor involvement in projects, it is important to consider the alternative means available to secure such involvement. These include:

- Binding contractual commitments under bespoke or standard form agreements, and a detailed review in Appendix B compares the treatment of preconstruction phase appointments under six standard form building contracts;
- Corporate integration of project team members under joint ventures, although there remains the need for the joint venture to conclude its separate contractual appointment by the client;
- Informal arrangements such as letters of intent which can govern limited preconstruction phase activities, but lack sufficient clarity to govern detailed arrangements;
- Non-binding or unwritten arrangements which may be used of necessity due to the speed of events or as a result of excessive faith in the goodwill generated by partnering being sufficient to govern complex preconstruction processes.

It is argued that the binding preconstruction phase agreements, preferably using published standard forms, offer the best means to achieve clarity of agreed preconstruction phase commitments, to avoid delay in completion of agreed activities and to avoid confusion of such activities with negotiation of the construction phase building contract.

## 11.6 Increased preconstruction commitments under framework agreements

Achievements derived from long-term multi-project relationships suggest that the commercial attraction of working on more than one project will have positive impact on investment in preconstruction phase processes. To support such relationships, a framework agreement can set out the common expectations of the parties and standard mechanisms for the detailed planning of individual projects.

Empirical research from three project case studies presents the benefits that can be obtained through the creation and use of properly structured framework agreements.

Framework agreements have a role in setting standard preconstruction phase processes linked to the implementation of successive projects. However, a review in Appendix C of two published standard form framework agreements suggests that neither contains provisions to clarify and control such preconstruction phase processes.

It is suggested that a successful framework agreement requires implementation of agreed processes and mutual commitments so as to establish new behaviour such as shared information, and a further project case study illustrates the problems that arise if this does not occur.

## 11.7 The relationship of preconstruction phase agreements to project management and partnering

Recognising that preconstruction phase processes often require independent project management, the appointed project manager will be in a position to influence the client in deciding whether its procurement strategy should provide for early contractor involvement. However, the credibility of the project manager's recommendations depends on its objectivity and it needs a medium to demonstrate this through shared information. It is suggested that early project planning and team integration undertaken in accordance with contractual programmes forming part of preconstruction phase agreements should assist project managers in doing their job.

Turning to partnering, this is distinguished from teamwork and is considered as a form of project management and planning. The following observations are relevant to the successful implementation of partnering:

- The importance of clear project processes and the risk of over-reliance on behavioural change by parties not familiar with project processes;
- The limited opportunity for changes in business culture and the possibility of achieving collaboration for specific agreed commercial purposes without changing the business culture of the collaborators;
- The challenges caused by changing business conditions, uneven levels of commitment, lack of momentum and failure to share information, and the role of the preconstruction phase agreement in overcoming these challenges.

The treatment of partnering in standard form building contracts is often limited to contractual declarations of partnering values, but such declarations need to be substantiated by underlying contractual processes through which they can be put into practice. Arguments that partnering may confuse contractual relationships by creating obligations of good faith and fiduciary relations, estoppel and waiver, confidentiality and disclosure can be neutralised by sufficiently clear contract terms.

## 11.8 Circumstances and attitudes that are obstacles to early contractor appointments or to preconstruction phase agreements

Preconstruction phase agreements are not as widely used as their potential benefits suggest they should be, and it is important to explore the possible reasons for this.

Certain projects are unlikely to benefit from joint preconstruction phase activities, such as small projects or arm's length design and build projects. Single-stage tendering remains popular irrespective of its shortcomings. Also, funding constraints may dictate a procurement strategy that focuses on financial close, although there may remain scope to integrate this with joint preconstruction phase activities in the context of long-term relationships.

Certain projects are subject to procedural and regulatory constraints. Also, the time and cost taken to create new agreements and to integrate the various agreements of different team members may be off-putting.

Negative attitudes are relevant and can include cynicism, aversion to contracts, conservatism and a wish to exploit a lack of contractual clarity. There may also be concerns as to the innate bureaucracy of contractual process, as to the relevance of contracts to personal relationships and as to whether contracts (in particular partnering contracts) involve dilution of legal rights. To address certain of these issues, there is the need for further industry education and training, but the appetite for such education and training is likely to be directly proportionate to the related prospects of new and profitable work.

## 11.9 Government and industry support for early contractor appointments under preconstruction phase agreements

It is recognised that to establish a new approach to procurement and new building contracts requires sustained support among clients and contractors.

There can be identified growing client support for early contractor involvement in successive government recommendations and construction industry reports. This appears unequivocal both as to the benefits of joint preconstruction phase processes and as to the value of preconstruction phase agreements to describe those processes. Against this background there is also evidence of the increased use of and support for contractor design contributions, two-stage pricing, selecting contractors by value, joint risk management, greater client involvement and partnering.

A substantial body of evidence reported from public and private sector case studies demonstrates the success of projects procured using early contractor involvement on individual projects and under framework agreements. Such projects compare well to the remainder of the industry in measures that include client satisfaction, defects, safety, cost predictability, time predictability, productivity and time on site. In addition, examples illustrate the specific experience of two major contractors as to increased use of preconstruction phase agreements and the related benefits to their businesses.

I hope that this book will assist in demonstrating the potential value of conditional preconstruction phase agreements as a means to secure greater benefits from early contractor involvement in project processes and as a means to underpin a partnering approach to project management that delivers tangible improved results.

So, has the case been proven to everyone's satisfaction and are the construction industry and their clients fully on board? Sir John Egan was asked to review performance ten years after *Rethinking Construction* and awarded us all four out of ten[748]. On the plus side, he noted demonstration projects completed at 20–30% less cost, taking 40% less time and delivery, 4% more profit for the industry – 78% of them without loss of time due to accidents as part of an 80% improvement in productivity.

However, on the minus side, Egan criticised relatively poor performance in particular sectors, such as private house-building which he said failed to achieve significant cost savings or productivity improvements. He also lambasted the Government for continuing to allow its own projects to be procured by single-stage lowest price tendering which he described as 'absolutely ridiculous'[749]. Egan emphasised that to achieve savings the client needs to understand cost, and that this requires the contributions of the whole team – 'a designer, a construction team, a supply chain and so on'[750], an argument squarely in favour of properly structured early contractor involvement. Egan may have been unfairly negative in his marking of the construction industry, given the recorded improvements in its performance on so many fronts. However, whether we see a period of sustained reform in construction procurement as a golden age or not, Egan's point is that there remains much more work that can be done.

---

[748] Egan (2008).
[749] Egan (2008).
[750] Egan (2008).

# APPENDIX A
# PROJECT CASE STUDIES

*Use of preconstruction phase agreements on single projects*

**Project case study 1**

> Bewick Court, Newcastle upon Tyne – refurbishment of high-rise residential block
> £3.5m refurbishment of landmark tower block in Newcastle upon Tyne, including major specialist recladding package

> Client: NBH, part of Places for People Group
> Main Contractor: Kendal Cross Holdings Limited
> Project Manager: Elliott Associates
> Architect: Red Box Design Group
> Engineers: Gilwood Engineering Services/BES Consulting Engineers/WSP

**Summary**

> Early appointment of main contractor – joint selection of cladding subcontractor and early appointment to contribute to design solutions – insolvency of cladding subcontractor – early warning and core group review – agreed actions to minimise delay and additional cost – client suspension of work – early warning and core group review – agreed actions to minimise additional cost.

**Key issues and events**

> On a £3.5m tower block refurbishment in the north-east of England, the housing association client with its project manager selected and

appointed a team comprising an architect, a structural engineer, a mechanical and electrical engineer and main contractor under a two-stage procurement approach. They signed a preconstruction phase agreement in the form of a PPC2000 Project Partnering Agreement pursuant to which they undertook design development, risk management and build-up of supply chain and cost information. This included the joint selection of a cladding specialist subcontractor whose package accounted for a significant element of the project (£1.5 m).

After start on site in late 2001, the cladding specialist (Allscott) went into administrative receivership. As a domestic subcontractor, the specialist's replacement was not the client's or consultants' contractual responsibility, and the main contractor could have been left simply to find a replacement at its own cost with full liability for any delay. However the project manager was aware, from the original joint client/main contractor tender undertaken to select the cladding specialist, that there were no comparable cladding specialists available within a wide geographical radius, and by serving an 'early warning' notice under PPC2000, invited the main contractor to put forward any alternative proposals at a 'core group' meeting of key individuals representing all team members.

The main contractor confirmed that engagement of an alternative cladding specialist would give rise to twelve (12) weeks delay and would cost the main contractor £175,000 – wiping out its profit and thereby its motivation to deliver a successful project. The main contractor's proposed solution was to take on direct liability for cladding, recruiting some of the insolvent specialist's workforce and buying the required materials cheap from the administrative receiver. This solution would involve a four week delay and additional costs of £7352. It was approved by the client.

The client had been reluctant to attend the required core group meetings, but was reminded that it was important to attend such meetings because PPC2000 provided that in the client's absence (or that of any other core group member) a binding decision could still be made by unanimous agreement of all core group members present. It was acknowledged that the agreed solution would be very unlikely to have been achieved if, for example, a JCT contract had been used, because:

- A JCT contract lacks a forum equivalent to the core group with an obligation to attend meetings and clear terms of reference for reviewing problems and looking for constructive solutions; and
- Under a single-stage approach to procurement leading to creation of a construction phase JCT contract the client would be unlikely to have full open information regarding pricing of the original specialist package distinct from the main contractor's profit and overheads.

*Appendix A*

The project team agreed that an important feature in overcoming this significant crisis was the role of the core group, who understood their functions under the partnering contract and who used consultation based on open-book price information to agree a compromise of strict legal entitlements that offered a major benefit to the project.

Later in the project, when three mobile phone companies delayed shutting down their aerials on the roof of the building, the project manager gave early warning of the need for suspension of work and requested the main contractor to propose a solution to the core group. The main contractor agreed to identify savings elsewhere in the project that could fund the costs of suspension (and those of its subcontractors) if any compensation received from the mobile phone companies was spent on the project. The client recovered £40,000 of compensation and spent it on an improved window-cleaning system.

## Contribution of preconstruction phase agreement

The success of the project team in dealing with the above problems is attributable to:

- The early appointment of the team, including the main contractor, under a preconstruction phase agreement supported by a key dates schedule setting deadlines for all preconstruction phase activities;
- The involvement of the client and its project manager with the main contractor in the early selection and appointment of the cladding specialist, as a result of which they had clear cost information (separate from the main contractor's profit and overhead) with which to analyse the cost consequences of replacing that specialist;
- The early establishment of a communications strategy, as a result of which in response to early warning the client (albeit reluctantly) agreed to participate with other partnering team members in meetings to consider solutions that went beyond strict application of the contract, and thereby enabled the team to react collectively to unforeseeable events;
- A clearly established role for the client and the project manager, as a result of which the client acknowledged that it should participate in core group activities and the project manager had the confidence to lead the client and other team members towards core group solutions;
- Creation by the above means of an opportunity for the main contractor to put forward an imaginative solution to a serious problem on site, with the confidence to take a commercial risk in assuming responsibility for the cladding specialist package, thereby effectively managing the risk arising from the subcontractor insolvency and

avoiding the prospect of a loss-making project and the likelihood of claims and cross-claims including those arising from the later suspension of work.

To quote David Pearson of Elliott Associates, project manager of the Bewick Court project:

'We could have seen contractual claims against both the client and the contractor and worst of all a project not yet concluded, resulting in another cold winter for Bewick Court residents.

Instead, the project finished on time and within its maximum price and the team remains firmly on speaking terms'[751].

**Sources**

- Review of contract and related documents and project correspondence;
- Reports and presentations prepared by Elliott Associates.

---

[751] Email to the author dated 31 October 2006

*Appendix A*

# Project case study 2

Watergate School, Lewisham – newbuild special school
£5m newbuild special school including hydrotherapy pool and other specialist equipment

| | |
|---|---|
| Client: | London Borough of Lewisham |
| Main Contractor: | Willmott Dixon Construction Ltd |
| Project Manager: | Technical Services Department, London Borough of Lewisham |
| Architect/Engineer: | FM Modern Design |
| Cost Consultant: | Potter Raper |

## Summary

Early appointment of main contractor – selection by value including alternative main contractor design solution – agreement of lump sum main contractor profit and overheads and lump sum design fees – use of preconstruction phase programme to meet deadlines for design development and selection of subcontractors and suppliers – specialist design contributions in relation to mechanical and electrical works and hydrotherapy pool – subcontractor and supplier prices in excess of cost plan – use of value engineering to bring costs within budget and maintain quality – core group including head teacher to ensure client involvement in decisions.

## Key issues and events

In autumn 2000, London Borough of Lewisham went out to tender for the design and construction of the new Watergate Special School. They issued designs prepared by their own architect and invited bidding main contractors to offer alternative design solutions. The criteria for the main contractor's selection were as follows:

- Innovative design solutions;
- Proposals for working with stakeholders;
- Proposals for managing supply chains;
- Proposals for site welfare and minimising accidents;
- Proposals for quality control and efficient rectification of defects;
- Lump sum design fees;
- Lump sum profit and central office overheads.

Willmott Dixon, working with FM Modern Design as architects and engineers, offered an alternative design solution, having visited the site and consulted with the school teaching staff. Willmott Dixon's tendered design solution was accepted by London Borough of Lewisham whose original architects were stood down.

London Borough of Lewisham wanted to adopt a partnering approach to achieve joint design development and finalisation of prices. The client, project manager, main contractor, design consultants and cost consultant attended several workshops together with the head teacher of the school and commenced a collaborative process of design development. Willmott Dixon was able to obtain early informal design contributions from prospective specialists seeking to be appointed in respect of mechanical and electrical works and the hydrotherapy pool.

However, the team was slow to agree a timetable governing preconstruction phase activities. In particular, FM Modern Design were reluctant to agree fixed dates for production of detailed drawings sufficient for Willmott Dixon to obtain fixed price subcontract and supply packages. Agreement of a binding timetable governing preconstruction phase activities, as part of a PPC2000 Project Partnering Agreement, was only concluded when it was illustrated that without this both the preconstruction phase and construction phase of the project were in danger of overrunning. It was acknowledged by the team that until that point the parties' enthusiasm for collaborative working, particularly the added motivation that came from their increasing awareness of the plight of some of the pupils at the special school, had led them to focus only on the excellence of their designs and not on the deadlines for commencement and completion of construction.

When detailed designs were produced and prices obtained, some of these prices for specialist items were higher than the amounts originally allowed for in cost plans proposed by the London Borough of Lewisham's cost consultant. This gave rise to concerns regarding Willmott Dixon's subcontract and supply tender procedures and required high level intervention to emphasise to London Borough of Lewisham:

- That Willmott Dixon took no benefit from increased costs, as it would receive lump sum amounts of profit and central office overheads regardless of any increase in the price of the project; and
- That Willmott Dixon and FM Modern Design had a contractual obligation under PPC2000 to value engineer designs so as to achieve prices within the London Borough of Lewisham budget, without incurring additional fees for whatever redesign work was necessary to achieve this.

*Appendix A*

The project proceeded and was completed within budget and was recognised by the Department for Education and Skills as a model special school.

## Contribution of preconstruction phase agreement

The early selection of the main contractor and its design team, and the use of a preconstruction phase agreement, assisted in achieving the following:

- The offer of alternative designs by the main contractor, sufficiently early in the project for the client to adopt these in place of its previously proposed designs, and with the benefit of sufficient detailed information from the main contractor's proposed architect/engineer to enable the client properly to analyse the main contractor's alternative designs prior to accepting them;
- The use of the core group to identify the need for and to agree a binding preconstruction phase timetable so as to ensure compliance with agreed deadlines for consultant design outputs and client design approvals;
- Involvement in project planning of project end users (teaching staff), thus assisting the team to obtain users' recognition of the design and cost limits available for their project and their acceptance of necessary compromises;
- Recovery from apparent cost overruns (as a result of price inflation compared to the client's original cost plan) using value engineering so as to achieve a price within budget and obtain client approval and consultant/main contractor commitment ahead of start on site to the price of a project that was fully designed and costed.

It is acknowledged that the commercial motivation of the main contractor in speculating on alternative designs at the point of tender, and in committing (with its architect/engineer) to undertake value engineering without additional fee recovery, was in part due to their wish to demonstrate achievement of a successful project in the education sector so as to establish a reputation that would help them win additional similar projects.

## Sources

- Review of contract and related documents and project correspondence;
- Report of Sir Michael Latham, Deputy Chairman of Willmott Dixon, to PPC2000 National User Group Conference, 5 October 2006;
- Discussions with FM Modern Design.

**Project case study 3**

Poole Hospital, Dorset – refurbishment and extension of operating theatres
£2.5 million refurbishment of six operating theatres and addition of a seventh

| | |
|---|---|
| Client: | Poole Hospital NHS Trust |
| Main Contractor: | Mansell Construction Services Limited |
| Project Manager: | McNaughts |
| Architect: | QP Architecture |
| Engineers: | Whicheloe MacFarlane MDP, Anthony Ward Partnerships |
| Mechanical & Electrical Specialist Subcontractor: | Lorne Stewart Plc |

**Summary**

Early main contractor appointment to contribute to full design and costing prior to start on site – main contractor and subcontractor contributions to mechanical and electrical designs – joint client/consultant/main contractor risk assessment of likely interruptions to work due to clinical needs – agreed identification of other work to progress during interruptions – client/consultant/main contractor/subcontractor agreement of construction phase programme to allow the client a continuous full operating schedule.

**Key issues and events**

The client needed to refurbish six operating theatres and add a seventh within twenty-four months, while enabling its hospital to maintain a full operating schedule. It also needed its main contractor to be willing and able to stop work immediately in the event of an emergency, as noise and vibration would not be permitted in the vicinity of the operating theatres. The client opted for a partnering approach using the PPC2000 form of contract as it considered that a conventional design and build contract, whereby all risks (including the need to stop work at short notice) would pass to the main contractor, would be unfair on prospective main contractors and would inflate their tender prices.

The process of establishing a preconstruction phase agreement and a supporting timetable of key dates enabled the parties to agree in advance other less urgent works that could be carried out on site at

*Appendix A*

those times when work on the main theatres had to stop at short notice due to unacceptable noise levels interfering with clinical activity. The timetable also allowed for phased possession to enable the client to retain maximum control of the site.

The agreed preconstruction phase activities were also crucial in enabling the main contractor and its mechanical and electrical specialist subcontractors to undertake detailed design work in conjunction with the design consultants, so that the project could be fully designed and costed before the agreed price was finalised, ensuring that the client did not pay for unnecessary main contractor risk assumptions.

The project manager met daily with clinical teams and the main contractor to ensure the works fitted around the clinical programme. The project completed on time and within budget and had no adverse impact on clinical performance.

The client subsequently used preconstruction phase agreements as part of its procurement strategy for further projects including:

- Breast screening unit (value approximately £2 million);
- Ward refurbishment (value approximately £4.8 million).

## Contribution of preconstruction phase agreement

The early appointment of the main contractor under a preconstruction phase agreement assisted the team in achieving:

- Main contractor and specialist mechanical and electrical subcontractor design input prior to start on site, so as to ensure that a fully designed project could be implemented, with any further design releases not causing delays or misunderstandings;
- Organisation of a phased project in a fully operational building, requiring a construction phase programme to be jointly prepared and agreed by the client, project manager, design consultants, main contractor and mechanical and electrical specialist subcontractor;
- Agreement among the project team of the client's clinical priorities, the need to manage time alongside site conditions and the risk of interruptions, and the agreement of a cooperative approach so as to minimise main contractor risk premiums;
- Implementation of a complex construction programme, with involvement of end user clinical teams to ensure integration with their clinical programme.

Clive Radestock, Capital Developments Manager, Poole Hospital NHS Trust, stated that: 'Having a mutual contract in place creates peer pressure to those team members with whom performance may be lacking, as the agreement is with each other rather than the client or contractor'.

He also stated: 'The process of risk management should be an intrinsic part of the project development, with clear lines of responsibility and required action to eliminate and/or mitigate the risk'[752].

**Sources**

- Review of contract and related documents and project correspondence;
- Discussions and correspondence with McNaughts and Poole Hospital NHS Trust.

---

[752] Email to the author dated 30 August 2005.

*Appendix A*

**Project case study 4**

>Bermondsey Academy, Southwark – newbuild school
>£22m newbuild academy school on restricted site, incorporating teaching and sports facilities

>| | |
>|---|---|
>| Client: | City of London Academy (Southwark) Ltd |
>| Main Contractor: | Willmott Dixon Construction Ltd |
>| Project Manager: | Department of Technical Services, City of London Corporation |
>| Designers: | Studio E (Architect), Dewhurst MacFarlane (Structural Engineer), Max Fordham (Services Engineer), FIRA Landscape (Landscape Architect) |

**Summary**

>Early simultaneous appointment of design consultants and main contractor – clarification of roles and responsibilities in preconstruction phase agreement and programme – joint management of cost/time impact of requirement for major changes in temporary school site – joint assessment and management of risk using risk register – use of core group to resolve differences – agreement of incentives to motivate cost savings and added value.

**Key issues and events**

>The City of London Academy (Southwark) Limited, financed by the Department for Education and Skills and the City of London Corporation, used the creation of a preconstruction phase agreement to clarify the roles and responsibilities of all team members, to progress design and risk management, and to ensure that the project could be built within the available budget. The client was keen not to lose its relationship with the design team, which it considered an inherent risk of the design and build approach and preferred to create its team through multi-party contractual obligations, including a preconstruction phase agreement, under PPC2000.
>
>The main contractor and design consultants were appointed simultaneously. They worked under an agreed timetable of preconstruction activities to develop innovative designs (creating 90% natural lighting and ventilation) that were affordable and achievable.
>
>The team also managed significant risks such as removal of on-site contamination, dealing with the presence of Japanese knotweed and

most significantly the last minute need to change the site of the required temporary school.

A key lesson from this project for the City of London Corporation was that any grey areas in the brief can lead to the other parties being reluctant to sign the preconstruction phase agreement. Certain consultants' insurers took time to be satisfied that multi-party preconstruction phase commitments did not create obligations beyond their clients' professional expertise.

It was also necessary to keep a close watch on the development of the preconstruction phase risk register, so as to ensure that risk contingencies allocated to particular items were subject to scrutiny and agreed actions by project team members, and so as to reduce or eliminate such contingencies wherever possible prior to finalisation of the agreed price for the project.

The project received praise from a number of sources, including the status of an exemplar project (Department for Education and Skills), Medium-Sized Project of the Year Award 2006 (Department for Trade and Industry and Constructing Excellence) and recognition of its environmental achievements by Jonathan Porritt (speaking at an open day at the Academy in October 2005). The project was also awarded the Prime Minister's Better Public Building Award 2006, the citation for which stated 'Partnering between the contractor and design team produced a strong building which benefited from nearly two years of preparation.'

## Contribution of preconstruction phase agreement

The use of a preconstruction phase agreement assisted in achieving:

- Understanding by all team members of their respective roles and responsibilities, particularly the interface of the design consultants among themselves and with the main contractor;
- Integration of preconstruction phase design development with finalisation of open-book prices so as to achieve a robust fixed price ahead of start on site, supported by detailed information to assist subsequent change and risk management;
- Successful management of risk of on-site contamination and the risk of switching sites for the required temporary school during the preconstruction phase;
- Agreement of performance measures so as to focus the team on the client's priorities;
- Early agreement of commercial incentives by way of shared savings so as to motivate the search for improved design and risk solutions.

*Appendix A*

The City of London Corporation has since adopted the use of preconstruction phase agreements in its procurement of two additional academy projects in Hackney and Islington, and also in its procurement of a large retail and office development in the West End of London.

It is interesting to note that on its subsequent academies and commercial project the City of London Corporation has entered into pre-construction phase agreements at an even earlier stage, using them to ensure that the design consultants and main contractor undertake joint feasibility studies and obtain the required planning consents before being authorised to proceed with detailed design and finalisation of prices.

John Frankiewicz, Chief Operating Officer of Willmott Dixon, stated that 'When commissioned at an early stage we will involve … ground work contractors who may be able to identify potential risks that could be avoided through consideration in regard to orientation of the building or the location/availability of drainage services'.

**Sources**

- Review of contract and related documents and project correspondence;
- Review of DVD prepared by Einstein Network for Royal Institution of Chartered Surveyors 2006;
- Review of additional materials provided by Willmott Dixon, by Construction News and by CABE;
- Correspondence and discussions with City of London Corporation and Willmott Dixon.

**Project case study 5**

Macclesfield Station – railway station refurbishment
£500,000 refurbishment of Macclesfield Station including the design and construction of new platform canopies

Client:                West Coast Trains/Virgin Trains
Main Contractor:       C. Spencer Limited
Project Manager:       Heery International
Design Consultants:    Atkins
Cost Consultant:       Faithful & Gould

**Summary**

Early appointment of main contractor for extensive preconstruction phase preparations – joint design development by consultant and main contractor including required third party approvals – joint risk management by client, consultant and main contractor including consultation with regulators and other stakeholders – agreement of fixed price with minimal main contractor risk premiums – agreement of construction phase programme in line with very restricted weekend closures.

**Key issues and events**

West Coast Trains/Virgin Trains had only a very limited time permitted by the Strategic Rail Authority for the refurbishment of Macclesfield Station, namely forty days of temporary closures over weekends. This led them to establish a preconstruction phase agreement so as to ensure that they (together with their main contractor, project manager and design consultants) could establish a reliable construction phase timetable that would fit within these tight constraints.

West Coast Trains/Virgin Trains had experience of projects overrunning under construction phase JCT contracts and also of main contractors charging substantial premiums for meeting time constraints. Hence, they wanted to use a preconstruction phase partnering approach as an alternative.

West Coast Trains/Virgin Trains were particularly conscious of the wide range of stakeholders who could influence the timing of preconstruction activities and construction activities. They used a preconstruction phase agreement under PPC2000 with a preconstruction

## Appendix A

phase programme to ensure that deadlines for all planning and design stages were agreed and committed to by project team members, and that the dates for all on-site activities were visible to and approved by Network Rail, the Strategic Rail Authority and the station manager.

The project involved a seven-month preconstruction phase (September 2002 to May 2003) and a five-month construction phase (to September 2003) and was completed on time and within budget.

### Contribution of preconstruction phase agreement

The preconstruction phase agreement assisted the project team in achieving:

- A fully designed and a costed project in advance of start on site;
- Testing of design assumptions by the main contractor as to their buildability and as to the constraints of the construction phase programme;
- Joint analysis and management of the risks of a tight construction phase so as to minimise main contractor risk premiums;
- Maximum involvement of third party stakeholders in approving preconstruction outputs and in agreeing an appropriate approach to the construction phase;
- Establishment of a clear communications strategy and a culture of trust and cooperation so as to ensure that the team could overcome problems on site without, for example, provoking main contractor claims for acceleration.

West Coast Trains/Virgin Trains agreed that this procurement model and contractual structure should be used on other refurbishment projects forming part of the redevelopment of its network of stations.

Kevin MacConville of Heery International, project manager on the Macclesfield Project, stated that:

> 'Preconstruction agreements either stand alone or as part of PPC2000 can be very effective provided such are utilised on the appropriate project, large-scope, complexity, etc.
>
> Employment of such an approach ... from the initiation stage of the project can/should increase the effectiveness and efficiency of the project team including client decision making process. The total process should provide improved value for money through focusing the team's attention to implement and hence achieve the activities as listed above.'
>
> Risk management is 'a significant area where open and frank discussions and workshops can really drive and control the project objectives and ensure value for investment is achieved ... The client has a most important role to play in this process – he needs to identify

and explain what risks he is willing to undertake and for how long'[753].

**Sources**
- Review of contract and related documents and project correspondence;
- Correspondence and discussions with West Coast Trains/Virgin Trains and Heery International.

---

[753] Email to the author dated 14 November 2005.

*Appendix A*

**Project case study 6**

Nightingale Estate, Hackney – housing refurbishment
£13m refurbishment of Rogate House, Nightingale Estate, Hackney including specialist concrete works and the cut-through of an existing building to form two separate buildings

Client:                London Borough of Hackney
Main Contractor:       Wates Construction Limited
Project Manager:       London Borough of Hackney/Leonard Stace
Design Consultants:    Abbey Holford Rowe (Architect), Babtie Allott (Structural Engineer), Lomax Consulting Engineers (Services Engineer)

**Summary**

Early appointment of main contractor – joint design development – specialist subcontractor contributions to specifications and to methodology for concrete repairs and major structural alterations – joint risk management including resident consultation and relocation – agreement of complex contractual construction phase programme including work to occupied properties.

**Key issues and events**

London Borough of Hackney had previously refurbished Alma House on the same estate in 1997 utilising a single-stage procurement model, but had underestimated the amount of asbestos in the building and the extent of deterioration of the concrete frame. These factors and lack of resident cooperation led to substantial delays and cost overruns. Also, defects discovered after completion led to ongoing problems with residents.

London Borough of Hackney wished to explore an alternative procurement approach that recognised the complexities of a major refurbishment project and enabled it to achieve time and cost certainty. For this purpose, it commissioned a bespoke two-stage partnering contract (which was an early prototype for PPC2000) and negotiated conclusion of this contract with its main contractor, project manager and design consultants.

The project team entered into a bespoke preconstruction phase agreement which established the joint basis for development of designs and risk management, with early appointment of specialist subcontractors, including concrete specialists, to analyse design solutions, methods of working and programmes.

The preconstruction phase ran from April to September 2000 and was followed by a ninety-week construction phase.

Prior to commencement of construction, comprehensive surveys were undertaken by the concrete and asbestos specialists. In addition, the entire team agreed the construction phase timetable in advance so that this could be the basis of consultation with residents and the establishment of a decanting programme.

Multi-skilled specialists were selected for internal works, in preference to traditional trades, so as to speed up the project and require fewer visits to each flat. This was instrumental in obtaining support from residents that was not evident on the earlier Alma House project.

All project team members were involved in detailed tenant consultations including tenant choice. This led, for example, to an agreement to refurbish rather than replace the roof, with consequent savings spent on improved quality kitchens and aluminium/timber windows plus ground level porches. All these variations were achieved within the project budget and agreed construction phase programme.

The collaborative approach to design also allowed agreement of aesthetic improvements such as external metal balconies which were designed and installed in collaboration with the balcony supplier for a cost less than that incurred at Alma House.

Another example of design collaboration involved the main contractor recommending metal refuse chutes as a cheaper and simpler alternative to the Alma House precast concrete chutes.

A major challenge for the project was the cut-through in the middle of Rogate House to create two separate blocks. Some residents refused to move because of problems in offering temporary accommodation, leading to a twenty-week delay to the project that was reduced to six weeks through agreement by the project team to resequence other activities. It was acknowledged by the main contractor that under single-stage procurement this risk event would have been exploited by them as a major time/cost claim, whereas the consultative team-based culture established on Rogate House, through two-stage procurement and the medium of the core group, led to the joint agreement of changes so as to minimise their delay and cost effects.

Other examples of collaborative working included:

- Designing and programming the works so that residents' water supplies were cut off only once for a few hours;
- The main contractor integrating its work with the requirement of the gas supplier to renew its equipment in each flat, so that only one visit was required to each flat for this purpose.

*Appendix A*

Completion of the works at Alma House had taken 115 weeks to refurbish 108 flats, whereas at Rogate House it took 90 weeks to refurbish 192 flats. The Rogate House team had worked at approximately double the speed.

The original budget of Alma House was exceeded by 18% whereas the original budget at Rogate House was exceeded by less than 6%.

London Borough of Hackney has since used the same procurement approach as the basis for appointing groups of main contractors to deliver Phases 1 and 2 of its Decent Homes housing refurbishment programme, covering works of a value in excess of £250m.

## Contribution of preconstruction phase agreement

The use of a preconstruction phase agreement assisted London Borough of Hackney and its project team to achieve:

- Joint investigation and planning of complex refurbishment works so as to achieve an accurate and integrated construction phase timetable;
- Consultation with residents as end users so as to establish their needs and priorities and enlist their cooperation;
- Involvement of the main contractor and its specialist subcontractors in design improvements, value engineering and specialist designs, working in cooperation with design consultants;
- Agreement of open-book prices so as to provide cost information necessary to manage change and risk without dispute;
- Joint risk management so as to minimise cost and time effects.

To quote Noel Foley, client project manager on the Nightingale Estate project: 'The preconstruction phase agreement must be part of the business case review or gateway management process as projects proceed to site. This could then act as a check on rushing to site'[754].

Leonard Stace, Cost Consultant on the project, observed 'We needed concrete specialists on board early in case we had to rejig the programme: they would know whether these changes were achievable. A close-knit team with a common objective ... was the only way this approach could work'[755].

## Sources

- Review of contract and related documents and project correspondence;
- Review of Housing Forum Demonstration Project report;
- Correspondence and discussions with London Borough of Hackney and Wates Construction.

---

[754] Email to the author dated 9 August 2006.
[755] *Building Magazine*, 10 May 2002.

**Project case study 7**

A30 Bodmin/Indian Queens – new road construction
£43 m 11.5 km new dual carriageway to replace existing A30 plus six kilometres of associated side roads, five over-bridges and four under-pass structures with new junctions

| | |
|---|---|
| Client: | Highways Agency |
| Main Contractor: | Alfred McAlpine Limited |
| Design Consultant: | Scott Wilson Kirkpatrick |
| Specialist Earthworks Subcontractor: | Kerbline |

**Summary**

Early appointment of main contractor – joint design development including choice of route to take account of buildability and traffic flow – joint risk management including early archaeological investigations, compulsory purchase procedures, ecological strategy and access issues and preparation for public inquiry, significant reduction of preconstruction phase and construction phase duration.

**Key issues and events**

The Highways Agency introduced a strategy for early contractor involvement (ECI) so as to ensure that its main contractors, working with design professionals, could offer advice and experience as to materials, methods and buildability issues. In order to establish a clear structure for early contractor involvement, the Highways Agency introduced a preconstruction appointment document by adapting and developing a construction phase NEC2 (Option C) building contract.

Alfred McAlpine were appointed as main contractor on the Bodmin/Indian Queens project (budgeted at £43 m) with designs still outstanding and costs still to be agreed. Their selection was based on 70% qualitative criteria and 30% price criteria (i.e. their quoted profit, overhead and preconstruction phase fees).

The contractor was incentivised by the offer of a 30% share in savings against budget achieved as a result of their design input. Working with the Highways Agency and its design consultant Scott Wilson Kirkpatrick, McAlpine had 21 weeks to finalise design, resolve all traffic management issues and provide an initial target cost. By the end of 39 weeks, the Highways Agency needed to issue draft orders for the road, including notification to landowners for compulsory purchase as well as its environmental statement, so that all of this information could be taken forward to a public inquiry.

*Appendix A*

The preconstruction phase programme subdivided key dates for Phase 1A and Phase 1B activities. In respect of Phase 1A, it recorded 478 agreed preconstruction phase activities (comprising design development, preparation of the environmental statement and issue of draft orders) each with an agreed deadline. Phase 1B was a subset of Phase 1A recording 225 items relating to preparation for and attendance at the public inquiry and assistance during the decision period.

McAlpine resolved many issues with local landowners that, in its absence, would have been likely to go forward to the public inquiry. For example, an access track to local land was due to be severed by a new road junction, so McAlpine proposed instead a link to the local road coming off the junction's southerly roundabout. At another point, McAlpine agreed to build a number of walls up to 1.8 m high in order to allay concerns regarding noise.

Other contributions offered by the main contractor during the preconstruction phase included:

- Early construction of side roads coming off the main dual carriageway so that these could be used as temporary traffic diversions for the A30;
- Assistance with 12 temporary traffic management systems to allow traffic on the A30 to continue uninterrupted;
- Scheduling of construction activities so that material was used to the maximum extent on site with very little going to landfill;
- Replacement of a proposed viaduct with a steep-sided embankment built from reinforced earth.

McAlpine also proposed a number of environmental measures, including reunifying the two halves of Goss Moor (a National Nature Reserve) previously divided by the old A30, by diverting the route of the new dual carriageway to the north and then degrading the old A30 back to its sub-base.

Early archaeological investigations avoided the risk of later discoveries leading to delay in the construction phase. Also, informal agreements with landowners, allowing reptile fences to be put up early, meant that relocation of snakes and lizards could proceed in an orderly manner. This was commenced early in April 2005, as relocation is more difficult during the summer months, and was completed so as to avoid up to six months' slippage in the construction phase.

To quote Keith Titman of Alfred McAlpine:

'The time available at the front end to plan how we deal with the archaeology and ecology, plus sort out issues with local landowners, is one of the key benefits with ECI. In times gone passed[sic], we have been awarded design and build contracts with just ten weeks lead-in, which ends up putting pressure on the whole programme'[756].

## Contribution of preconstruction phase agreement

Use of a preconstruction phase agreement under the Highways Agency ECI procurement method assisted in:

- Obtaining main contractor design contributions as to the most appropriate route, the re-use of fill on site and the substitution of specialist earthworks for a viaduct;
- Collaborative working between the client, main contractor and design consultant to manage risks arising from the need for consultation with local landowners, completion of compulsory purchase and submission to a public inquiry;
- Efficient programming of preconstruction activities such as archaeological surveys and ecological measures so as to avoid delays in the preconstruction phase.

Bodmin/Indian Queens has been described as 'the first pure ECI' project (Keith Titman, Alfred McAlpine, *Contract Journal*, 26 July 2006) and is part of a wider Highways Agency strategy to implement numerous projects utilising the ECI procurement route.

In McAlpine's view the benefits of early contractor involvement included merging the skills of designer and contractor to deal with planning and buildability, design and statutory procedures, environmental assessments and cost estimation. This provided the client with 'a more robust design and price' and also facilitated the 'large time savings' achieved through parallel working[757].

Additional benefits identified by McAlpine included early supply chain involvement in the design process, so that designs could be developed to suit locally available materials and local skills.

## Sources

- Review of contract and related documents and project correspondence;
- Review of Highways Agency Procurement Strategy;
- Correspondence with Alfred McAlpine and review of Alfred McAlpine presentation material.

---

[756] McAlpine slays the beast of Bodmin. *Contract Journal*, 26 July 2006, pp. 24, 25.
[757] Email from Keith Titman to the author dated 22 December 2006.

*Appendix A*

**Project case study 8**

Project X – newbuild housing scheme with modular units
£8 m newbuild housing scheme in Central London utilising modular kitchen and bathroom pods
Client/main contractor/project manager and architect/structural engineer/cost consultant: Details confidential

**Summary**

Design conceived by specialist manufacturer – architect, engineer and main contractor all joined team later – creation of preconstruction phase agreement – primary focus of team on design issues – architect also project manager – problems during construction phase – underground obstructions – design details or variations – lack of client leadership – late notification of escalating claims for additional time/money – use of core group to negotiate solutions.

**Key issues and events**

The scheme was conceived by the specialist kitchen/bathroom pod manufacturer and its construction company parent, and the housing association client brought in an independent architect and structural engineer to complete the team. The architect was appointed as lead designer and was also appointed as project manager. All team members agreed to utilise a two-stage project partnering approach under the PPC2000 form of partnering contract. A preconstruction phase agreement was entered into and the team members proceeded with design development and the pricing of works packages.

It became clear that the main contractor parent of the modular supplier (being based in the north-east of England) was unable to obtain competitive prices for other work packages from supply chain members in London, and it was agreed that the main contractor role would be retendered while preserving the appointment of the modular supplier.

The substitute main contractor therefore joined a team that was already well established, working with designs that were well advanced and that centred around the concept previously put forward by the modular specialist supplier.

The senior representatives of each organisation named as core group members failed to meet on a regular basis or to build up efficient working relationships that could have assisted them in agreeing solutions when problems arose. In addition, contrary to the terms of the preconstruction phase agreement, the team failed in their construction phase key dates schedule to set out clearly the required stages for release of design information and the procedures governing the firming up of provisional sums. This left considerable scope for argument as to what information was required from the architect, from the engineer, from the main contractor and from the specialist modular supplier at various stages during the construction phase.

After an initial workshop, there was no further training or detailed consultation regarding the bespoke contract terms until problems started to arise during the construction phase.

Problems included discovery of a large drainage pipe crossing the site which had not been evident on survey. In addition, the site was constrained by its proximity to a busy railway line and a main road classified as a red route, giving rise to access and mobility problems only partly alleviated by the choice of modular units to minimise work on site.

The architect issued over 130 'architect's instructions' (not a recognised term under the agreed form of contract) and did not (as required by the form of contract) state whether or not these constituted changes. In turn, the main contractor did not (as required by the form of contract) provide advance submissions identifying the time and cost consequences of those instructions that it considered to be changes before implementing them on site. In addition, the client and its cost consultant took a passive role, attending meetings but not noticing or not questioning the other parties' failure to adhere to contract procedures – notwithstanding that elsewhere in London the same client was using the same form of contract with another team on another project.

When the main contractor submitted its accumulated claims, they amounted to over £1.5m and the client started to consider its position under the contract. Core group meetings were called, facilitated by the 'partnering adviser' appointed under the form of contract. Agreed representatives of all partnering team members were contractually obliged to attend core group meetings and the contract provided that core group decisions required to solve the problems could be made by consensus of those core group members present at each meeting.

Over a series of meetings, including one to which the main contractor brought a notice of adjudication (which it did not serve), the core group members step by step acknowledged the parties' respective contributions to the problems and ultimately agreed a compromise. This involved deductions from the architect's and engineer's fees, and a partial payment of the main contractor's claim met by additional client funding.

The team used the core group to settle their differences and the completed project was recognised as a success. However, the previous adoption of behaviour and practices that ignored the contractual pre-construction phase processes caused a major distraction.

*Appendix A*

**Reasons for problems encountered**

Despite the use of a preconstruction phase agreement, the problems on this scheme derived from:

- Failure of team members to understand the preconstruction phase agreement and the significance of the preconstruction phase activities;
- Failure of the core group members to establish efficient working relationships and build up mutual confidence;
- Failure during the preconstruction phase to agree a properly structured construction phase key dates schedule;
- Conflict of interest on the part of the architect, who, as project manager, was required to assess delays caused by its own late issue of design information;
- Failure of the architect as project manager to apply the contract procedures regarding instructions and changes;
- Failure by the main contractor to apply contract provisions requiring early warning of problems and advance evaluation of the time and cost consequences of changes;
- Failure by the client and cost consultant to take an active role as core group members.

To quote the Chief QS of the construction company involved:

'Two-stage tendering is now commonplace in design and build, but we still see projects with no structured approach to the preconstruction activities ... Experience has taught us either:

- The notional preconstruction period quoted is often significantly exceeded, leading to unrecoverable staff costs; or
- A bespoke agreement is introduced later, when we are already committed and it is difficult to withdraw.

Without [a detailed and properly structured preconstruction phase agreement]:

- Unnecessary delays may occur in the preconstruction activities;
- Consultants continue to hold the contractor at arm's length over the design;
- The contractor is often appointed too late to influence the design;
- Parties manage risks in isolation, rather than as a team.

As to where [preconstruction phase agreements] are inappropriate single-stage tenders and/or where the criterion for winning the work is the lowest cost. In the latter case, we would not want to use a pre-commencement agreement because it removes risk and therefore the opportunity to exploit changes and delays to improve the commercial return'[758].

**Sources**

- Review of contract and related documents and project correspondence;
- Correspondence and discussions with project team members.

---

[758] Email to the author dated 12 October 2005.

*Appendix A*

# Use of preconstruction phase agreements on multi-project frameworks

## Project case study 9

> Whitefriars, Coventry – housing refurbishment programme
> £230m programme of housing refurbishment in 19,700 homes over a five-year period

> | | |
> |---|---|
> | Client: | Whitefriars Housing Group |
> | Main Contractors: | Wates Construction Limited and Lovell Limited |
> | Specialist Contractors: | Graham Holmes Limited and Anglian Windows Limited (Windows Specialists) |
> | Project Manager: | Officers of Whitefriars Housing Group with support from FPD Savills |

## Summary

> Early appointment of specialist window subcontractors on pilot projects – early appointment of two main contractors under joint framework agreement and annual building contracts – joint design development and procurement of supply chain including value engineering and agreement of shared kitchen supplier – joint risk management including resident consultation and joint training and employment initiative – use of three-way core group and early warning system – agreement of joint targets and incentives for cost and time savings.

## Key issues and events

> Whitefriars Housing Group, established following a large-scale voluntary transfer of housing stock from Coventry City Council, set up a strategic partnering arrangement with window specialists Graham Holmes and Anglian Windows in late 2000, the first use of the published PPC2000 Project Partnering Contract.
>
> This programme was expanded to comprise a total of £230m of refurbishment work undertaken by Wates Construction and Lovell, in collaboration with Graham Holmes and Anglian Windows among other specialist subcontractors.
>
> A three-way framework agreement was entered into by Whitefriars with Wates and Lovell, establishing the conditions for the award of annual programmes of work according to available client funding, contractor performance on previous work and contractor capacity for further work. It was accepted that the volumes of work allocated to each contractor could vary.

When the framework agreement was set up, Whitefriars Housing Group did not have sufficient funding (under private finance from Nationwide Building Society) to cover its entire programme, and it was recognised by the team that efficiencies were essential in order to reduce anticipated costs.

First, it was noted that Lovell had obtained cheaper prices from its long-term kitchen supplier, and Wates agreed to utilise the same kitchen supplier, with all consequent savings reverting to the client.

In addition, the establishment of a steady volume of work enabled both main contractors to operate using a stable workforce and to increase their efficiency on site, for example reducing the turnaround time for the installation of new kitchens from three weeks to two weeks per flat. Reduced time on site achieved savings in preliminary costs of £2m, again all reverting to the client.

At the same time, the client with both main contractors and in partnership with Mowlem, established the Whitefriars Housing Plus Agency which secured training opportunities for 38 people in the first year of the programme and a total of over 200 during the programme as a whole.

Costs were reduced to such an extent that Whitefriars could fund its entire anticipated programme. The programme was completed within five years rather than the anticipated six years (a 20% saving in time), at costs that were 10% lower than those originally agreed.

## Contribution of framework agreement/preconstruction phase agreement

Benefits achieved through the use of a framework agreement, in conjunction with preconstruction phase agreements for each annual phase of work, included:

- Recognition by both main contractors of the client's budgetary limits and the need for reduced costs in order to complete the entire programme;
- Establishment of a forum between the client and the two main contractors for exchange of information and shared best practice, leading to use of the most economical common kitchen supplier;
- Agreement and implementation of a shared approach to training via the Housing Plus Agency;
- Regular consultation, through a three-way strategic core group under the framework agreement, to identify opportunities for improved efficiency leading to more rapid turnaround on site;
- Engagement of key specialist subcontractors, initially under direct client appointments of windows fabrication and installation specialists, and subsequently through subcontract frameworks entered into

by the main contractors with these specialists and with key suppliers of kitchens and bathrooms;
- Reduced client need for external consultant support once the agreed preconstruction phase and construction phase procedures became well-established between the client and its main contractors and specialist subcontractors.

The commercial motivation for Wates and Lovell was partly the client's use of key performance indicators by reference to which the contractors needed to achieve agreed targets in order to obtain the award of successive years' work. The contractors were also motivated in part by the substantial volume of comparable housing refurbishment work likely to become available from other similar clients, and the need to demonstrate good performance with Whitefriars Housing Group in order to improve their prospects of winning such further work.

To quote Dave Smith, Director of Wates Group Limited (one of the two main contractors on the Whitefriars programme): 'Supply chain includes the client. If there is regular business with continuous improvement the supply chain will work. Continuous foreseeable work is still the essential ingredient, with patience a key factor, but we must start repeating details regularly'[759].

**Sources**

- Review of contracts, related documents and project correspondence;
- Review of Housing Forum case study;
- Correspondence and discussions with Whitefriars, FPD Savills and Wates Construction.

---

[759] Email to the author dated 12 October 2005.

## Project case study 10

Job Centre Plus – national office refurbishment programme

£737m fast-track programme for Department of Work and Pensions and Land Securities Trillium requiring the refurbishment and creation of 969 integrated Job Centres and Benefits Offices through a national roll-out programme

| | |
|---|---|
| Client: | Department of Work and Pensions (DWP) and Land Securities Trillium (LST) |
| Main Contractors: | Mowlem plc, Wilmott Dixon Limited, B&K Building Services Limited, David Maclean Contractors Limited, Rok Build Limited, Shepherd Construction Limited, Interserve Project Services Limited, Mansell Construction Services Limited, Overbury plc, Curzon Interiors Limited, Midas Property Services (UK) Limited, Longcross Limited, Banner Holdings Limited, and Styles and Wood Limited |
| Project Managers: | Seconded from consultancies such as Turner & Townsend, Gleeds and Atkins, coordinated by LendLease Projects. |

### Summary

Early appointment of 14 main contractors by two clients under multi-party framework agreement and successive project-specific building contracts – early appointment of specialist subcontractors and suppliers under additional framework agreements and successive project-specific subcontracts – fast-track preconstruction phase procedure and programme for each project – joint design development by design consultants and main contractor selected for each project – joint risk management through shared information among both clients and all main contractors so as to improve project processes – open-book pricing linking out-turn costs to key performance indicators governing award of future projects.

### Key issues and events

DWP and LST jointly selected a total of 14 main contractors and a range of specialist subcontractors to undertake a nationwide programme for conversion of Job Centres and Benefits Offices to provide combined Job Centre Plus offices. This was a fast-track programme utilising standard designs, materials and equipment adapted to a wide variety of different buildings.

*Appendix A*

DWP's and LST's joint objective was to create an efficient contract structure to enable a quick start-up on site, utilising model processes and contract documents to streamline a nationwide programme. They subdivided England, Wales and Scotland into districts, and a main contractor was appointed to undertake works in each district. DWP and LST wanted to ensure that there was a cross-pollination between districts, which was initiated through the use of a single multi-party Framework Agreement between all 14 main contractors and the joint clients.

DWP and LST also wanted to create a fully integrated supply chain to support the roll out programme, with Specialist Framework Agreements negotiated in parallel with key subcontractors and suppliers. Strict timetables were agreed to govern both the preconstruction and construction phases of each project.

The Job Centre Plus programme met its objective of transforming the Job Centre and Social Security Network and, with a final cost in 2006 of £737 m against a 2003 forecast of £981 m, achieved savings of 24.8%.

It won the Building Magazine Integrating the Supply Chain Award 2004; the LABC Services Award for Integrated Site Safety 2004; the OGC Government Opportunities Award for Public Procurement Excellence 2003; and the Building Magazine Health and Safety Award 2003.

**Contribution of framework agreement/preconstruction phase agreement**

Achievements from the use of a Framework Agreement included:

- Creation of a 16 party framework covering the entire £737 m Job Centre Plus programme, agreed on the same terms with all 14 main contractors;
- Development of a standard preconstruction phase agreement for use on 969 projects over an intensive series of preconstruction phase processes;
- Integration of roles and responsibilities of a national team of seconded project managers;
- Provisions for adjusted workloads among contractors according to measured achievement of targets for cost, time, health and safety and environmental impact.

The form of preconstruction phase agreement was based on PPC2000, adapted for the Job Centre Plus programme. Use of a preconstruction phase agreement on each project helped to achieve the following:

- Agreed stages of design development by the joint clients' appointed consultants, signed off and costed by each appointed main contractor on each project prior to start on site;

*Early Contractor Involvement in Building Procurement*

- Advance agreement of a demanding timetable for the construction phase of each project;
- Identification, management and minimisation of risks arising on each site;
- Identification and achievement of significant savings;
- Successful health and safety preconstruction phase processes;
- Integration of the Office of Government Commerce Gateway Process with the timetable for mobilisation and risk management of each project.

**Sources**

- Review of contract related documents and project correspondence;
- Review of additional material provided by Lend Lease Projects;
- Correspondence and discussions with Lend Lease Projects;
- NAO 2008.

*Appendix A*

**Project case study 11**

Eden Project Phase 4, Cornwall – education and leisure development
Design and construction of the award-winning Education and Resource Centre (the 'Core') together with a range of additional buildings and infrastructure on the Eden Project site

| | |
|---|---|
| Client: | Eden Project Limited |
| Main Contractor: | McAlpine Joint Venture (Alfred McAlpine Construction and Sir Robert McAlpine) |
| Project Manager and Cost Manager: | Davis Langdon |
| Planning Supervisor: | Waterman Burrow Crocker |
| Design Manager and Supervisor: | Scott Wilson Kirkpatrick |
| Lead Design Consultant: | Nicholas Grimshaw & Partners |
| Services Engineer: | Buro Happold |
| Civil and Structural Engineer: | Anthony Hunt Associates |
| Architect: | Haskoll Limited |
| Landscape Architect: | Land Use Consultants |

**Summary**

Early appointment of main contractor joint venture under multi-party framework agreement with client, project manager/cost manager and design consultants – joint design development including buildability and affordability of innovative designs – joint risk management including continued operation of facilities during construction phase – agreement of target cost for construction phase with pain/gain shares.

**Key issues and events**

The Eden Project had already procured Phases 1, 2 and 3 comprising the 'biomes' and main infrastructure under NEC2, working successfully with the above team and driven by a strong client-led culture and the high profile of the Eden Project itself.

The client and project manager recognised that Phase 4 would comprise a large number of projects each with its own timeframe that must not interfere with the ongoing operation of the Eden Project as a visitor attraction. Accordingly, they established a new contractual framework under which:

- The client, the main contractor and all consultants signed a single multi-party Framework Agreement;
- The Framework Agreement provided for joint design development, joint supply chain/ price build-up, joint risk management and joint agreement of construction phase programmes for each of the projects comprising Phase 4;
- The Framework Agreement created a system for instructing surveys and other early activities necessary to finalise the scope, price and programme of each project;
- The Framework Agreement provided for a project group with agreed terms of reference and an express duty to warn each other of problems arising;
- Following successful completion of the preconstruction phase processes described in the Framework Agreement, separate NEC2 contracts (using Option C Target Cost) were created in respect of each project forming part of Phase 4;
- Under the NEC2 contracts, cost savings of between £100,000 and £800,000 below the target price were agreed to be shared equally between the client and main contractor.

**Contribution of framework agreement/preconstruction phase agreement**

The Framework Agreement allowed the design consultants to work in direct liaison with Eden Project as client, and then to become sub-consultants of the McAlpine Joint Venture during the construction phase of each project, supporting McAlpine Joint Venture's single point responsibility for design, supply and construction.

Particular preconstruction contributions made by McAlpine Joint Venture as main contractor included site investigations and enabling works, ensuring that the china clay pit on which the Eden Project is built was suitable for the new Education and Resource Centre structure and ensuring that the design consultants were made aware of the impact of ground conditions in a timely manner.

The Framework Agreement assisted the Eden Project to achieve successful completion of its Education and Resource Centre, known as 'the Core' and opened by HM The Queen in summer 2006, as well as successful completion of a wide variety of other less high profile projects, utilising the same team working according to a pre-agreed set of project processes.

*Appendix A*

**Sources**

- Review of contract related documents and project correspondence;
- Correspondence and discussions with Eden Project, Davis Langdon and Alfred McAlpine.

**Project case study 12**

Project Y – education programme
£300 m programme for newbuild and refurbishment works at schools
Client/main contractor/design consultants/project manager: Details confidential

**Summary**

Framework for programmes of schools projects – framework agreement incorporated preconstruction phase processes – rigorous main contractor and specialist subcontractor selection – changes of senior client staff at commencement of programmes – lack of training of project staff – lack of joint client/main contractor preconstruction phase activities – slow start to programmes.

**Key issues and events**

The client wished to select a main contractor to undertake a ten-year programme of newbuild and refurbishment schools projects. The main contractor was selected, together with key specialist subcontractors on the basis of a framework agreement and PPC2000 project contracts that incorporated preconstruction phase agreements.

The client invested in a rigorous selection process for its main contractor and specialist subcontractors. However, the client was then hit by changes in its senior staff at a time which coincided with commencement of the programme.

The consequent break in continuity of staff representing the client organisation led to a lack of committed and experienced client officers implementing both the framework agreement and the preconstruction phase agreements for each project. Opportunities were therefore lost to commence the programme with confidence.

The main contractor had resourced its programme based on an assumed level of work. When very few projects commenced during the first year, the main contractor reduced the quality and level of resources committed to the programme.

The framework included key performance indicators allowing the client to measure main contractor and specialist subcontractor performance, linked to incentives governing increased profit and extension of the framework programme. However, these systems were not implemented at an early stage, as a result of which the main contractor ceased to see the key performance indicators as important measures of performance.

Client officers, project managers and main contractor staff were allowed to implement projects independently without the benefit of

*Appendix A*

training in the new preconstruction phase agreements. As a result, they did not conclude preconstruction phase agreements in the manner intended and lost the opportunity to work jointly with each other and with specialist subcontractors on designs and thereby to seek cost savings by means of value engineering.

The client and its main contractor worked hard to implement a major programme of work, but the problems experienced in adopting efficient preconstruction phase systems were a barrier to getting the best out of the client's investment in the framework.

**Reasons for problems encountered**

The problems that arose on the above programme were due to:

- Lack of client leadership due to break in continuity of key personnel;
- Failure by the client to apply provisions of the framework agreement that would motivate main contractor performance through the use of key performance indicators and incentives;
- Lack of training and slow/disjointed implementation of individual projects, resulting in failure to apply the joint preparatory processes set out in preconstruction phase agreements.

**Sources**

- Review of contract and related documents and project correspondence;
- Correspondence and discussions with project team members.

# APPENDIX B
# PRECONSTRUCTION PHASE PROCESSES UNDER STANDARD FORM BUILDING CONTRACTS

## 1 Introduction

This appendix reviews the structure and relevant provisions of a number of published standard form building contracts in order to consider the extent to which they describe preconstruction phase processes and create, or could be adapted to create, preconstruction phase agreements. The processes reviewed are:

- design development
- two-stage pricing
- risk management
- communications
- programmes
- integration of the team

The standard forms reviewed are as follows, for the reasons stated:

(1) GC/Works/1 Two Stage Design and Build, published in 1999, was the first standard form to describe the preconstruction phase appointment of the main contractor.
(2) NEC3, published in 2005, includes a Professional Services Contract that can be adapted to create a preconstruction phase agreement. It provides for joint risk management and has a partnering option. NEC, in its first edition, was recommended, subject to proposed amendments, by Sir Michael Latham in his 1994 report *Constructing the Team*[760]. NEC3 has been recommended by the Office of Government Commerce as compliant with their guidelines for Achieving Excellence in Construction[761].

---

[760] Latham (1994) 5.17 to 5.20 inclusive.
[761] NEC3 carries the following endorsement: 'This edition of the NEC (NEC3) complies fully with the AEC principles. OGC recommends the use of NEC3 by public sector construction procurers on their construction projects.'

*Appendix B*

(3) PPC2000, published in 2000 and amended in 2008, is a multi-party standard form that is stated to be a partnering contract and includes a conditional preconstruction phase agreement describing a range of preconstruction phase processes. PPC2000 has been supported by Sir John Egan and Sir Michael Latham and endorsed by Constructing Excellence, the Construction Industry Council and the Housing Corporation[762].

(4) Perform 21, the Perform 21 Public Sector Partnering Contract, was published in 2004. It includes a multi-party partnering agreement and a preconstruction phase agreement designed to govern preparatory activities undertaken by main contractors, subcontractors and suppliers prior to start on site.

(5) JCT 2005: although the JCT 2005 suite does not provide for a conditional preconstruction phase agreement, the JCT 2005 MPCC form provides for a design development process. In addition, the JCT published in 2008 a Pre-Construction Services Agreement, the JCT PCSA[763], for use between a client and its main contractor, and a separate Pre-Construction Services Agreement (Specialist)[764], the JCT PSCA(SP), for use between a client or main contractor and a specialist contractor or subcontractor.

(6) JCT CE – the JCT Constructing Excellence Contract was published in 2006. It is structured as a universal 'purchase order' which can be adapted to form a preconstruction phase agreement.

# 2 *Design development*

## 2.1 GC/Works/1 design development

The GC/Works/1 Two Stage Design and Build form creates the conditional appointment of a main contractor to participate in the design process, converting to an unconditional appointment when designs, supply chain arrangements and prices are sufficiently detailed for the project to commence on site[765].

Significant features in the design process envisaged by GC/Works/1 Two Stage Design and Build include:

---

[762] Sir John Egan launched PPC2000 in September 2000 and described it as 'a blow for freedom', (PPC2000 launch at the Building Centre, Store Street, London W1 (September 2000)). Sir Michael Latham called it 'the full monty of partnering and modern best practice', Latham (2002).
[763] JCT PCSA.
[764] JCT PCSA(S).
[765] GC/Works/1 Two Stage Design and Build, clause 10B (Design Process and Contract Sum).

- The requirement for the main contractor to 'carry out any site and/or soil investigations required by the Contract'[766];
- The requirement for the main contractor to 'proceed with the Design so that the Design Process Event shall be achieved in accordance with the Programme' (the Design Process Event being 'completion of the Design', 'signing and dating of all Design Documents', 'determination of Contract Sum' and any other events specified in the Abstract of Particulars)[767];
- Prevention of the main contractor taking possession of the site until so agreed (in particular until the Contract Sum is agreed) and the ability of the client to decide not to proceed with the contract at any time up until the main contractor is entitled to take possession of the site[768];
- The requirement for the main contractor to submit in accordance with the Programme 'a written fully itemised lump sum quotation of the proposed Contract Sum', thus allowing the main contractor's design contributions to inform the build-up of the quoted prices[769].

Despite the preconstruction phase design process established by GC/Works/1 Two Stage Design and Build in 1999, this form does not appear to have been significantly used in practice[770].

The following year GC/Works introduced amendments to GC/Works/1 Two Stage Design and Build and its other standard form building contracts that impose more speculative design obligations on the part of the main contractor. These provide for main contractor participation in design development by requiring it to undertake value engineering appraisals throughout the design and construction of the project 'to identify the function of relevant building components, and to provide the necessary function reliably at the lowest possible cost in terms of whole life costing or improved functionality at the same whole life costs'[771]. The main contractor is entitled to submit value engineering reports proposing changes to the client which could include proposals for sharing consequent reduced costs.

However, it seems unlikely that in practice a main contractor would do the work required to undertake such value engineering and submit the outputs of that work before agreeing with the client the basis on which its efforts might be rewarded. This GC/Works provision fails to recognise that value engineering should involve the integration of

---

[766] GC/Works/1 Two Stage Design and Build, clause 10B(1).
[767] GC/Works/1 Two Stage Design and Build, clause 10B(1).
[768] GC/Works/1 Two Stage Design and Build, clause 34.
[769] GC/Works/1 Two Stage Design and Build, clause 10B(2).
[770] RICS 2001 records one instance of its use, RICS (2001), 35. RICS 2004 records no instances of its use, RICS (2004), 27.
[771] GC/Works/1 (2000), Value engineering (Condition 40 (PM's instructions)).

*Appendix B*

contributions by all design team members and be linked to a system of reward agreed in advance. This is necessary to motivate the main contractor's efforts and also to clarify in advance the impact of a value engineering proposal on consultants. For example, value engineering may give rise to the need for consultant redesign work and claims for additional fees that would need to be set-off against any saving.

## 2.2 NEC3 design development

NEC3 contains only basic provisions describing a design development process, namely a requirement for contractor designs to be submitted to the project manager for acceptance[772]. NEC3 allows for this process to be clarified if design submissions and acceptances are allotted 'key dates'[773] for their completion, although this does not extend to preconstruction phase design submissions which would need to be subject to a separate NEC3 Professional Services Contract.

## 2.3 PPC2000 design development

A preconstruction phase design process is set out in PPC2000, the structure of which provides for the main contractor and the design consultants to be appointed simultaneously under a single multi-party agreement that will then govern their joint development of designs for the project in agreed stages.

Significant features of the PPC2000 design development process include:

- Agreement of a lead designer and other design team members, which can include not only design consultants but also the main contractor and appropriate specialist subcontractors and suppliers, all of whom sign a single form of contract which sets out their respective design roles and responsibilities[774];
- Provision for a standard design development process which can be amended by agreement[775];
- Clarification that each team member is responsible for errors and omissions in documents that it prepares or to which it contributes to (i.e. that team members do not have responsibilities for each

---

[772] NEC3 core clause 21.2.
[773] NEC3 core clause 30.3.
[774] PPC2000 Project Partnering Agreement.
[775] PPC2000 Partnering Terms, clause 8.3 and corresponding provision in Project Partnering Agreement.

other's documents), except to the extent that it is stated in the contract documents that one team member has relied on contributions or information provided by others[776];
- A specific sequence of preconstruction phase design development activities commencing with outline designs and alternative design solutions, followed by development of designs for client approval at each stage and then finalisation of detailed designs sufficient to satisfy any planning requirements and other pre-commencement regulatory approvals[777];
- Agreed periods of time for each stage of design development to be set out in the preconstruction phase programme known as the 'Partnering Timetable'[778];
- Provision for the results of surveys and investigations to be reviewed by the design team and reflected in any required design amendments subject to prior client approval[779];
- The requirement that, at each stage of design development, the lead designer and other design team members take into account the agreed project budget and provide updated cost estimates reconciled with that budget[780];
- Provision for value engineering at each stage of design development subject to client approval of value engineering proposals[781];
- Provision for main contractor objection to any designs to which it has not contributed, if these are contrary to the contract documents or 'otherwise demonstrably not in the best interests of the Project'[782];
- Provision for designs to become contract documents once they have been approved by the client[783].

## 2.4 Perform 21 design development

Perform 21 does not contain provisions describing a preconstruction phase design development process in its Prestart Agreement[784]. In its construction phase building contracts, it refers only generally to submission of design 'for acceptance'[785].

---

[776] PPC2000 Partnering Terms, clause 2.4.
[777] PPC2000 Partnering Terms, clause 8.3.
[778] PPC2000 Partnering Terms, clause 8.3.
[779] PPC2000 Partnering Terms, clause 8.4.
[780] PPC2000 Partnering Terms, clause 8.7.
[781] PPC2000 Partnering Terms, clause 8.8.
[782] PPC2000 Partnering Terms, clause 8.11.
[783] PPC2000 Partnering Terms, clause 8.12.
[784] Perform 21, PSPC 10.
[785] For example, Perform 21 PSPC 3, clause 6.2.

*Appendix B*

## 2.5 JCT 2005 design development

Most JCT 2005 contracts provide for contractor design contributions only during the construction phase of the project, for example the provision for a Contractor's Designed Portion in JCT 2005[786].

JCT 2005 MPCC includes a design submission procedure and envisages that this will be implemented by reference to an agreed 'design programme'[787]. The extent to which this provision could be utilised during the preconstruction phase is limited, as the provisions for commencement on site[788] envisage a set date without any apparent conditionality (i.e. without pre-conditions dependent on preconstruction phase design development). Nevertheless, the JCT 2005 MPCC Guidance Notes mention specifically that access to the site may be given 'some time after commencement of the Project in order to allow time for design and other preparatory activities to take place'[789].

JCT 2005 MPCC assumes a 'design and build' responsibility on the part of the main contractor and does not allow for design submissions to be made by or on behalf of the client and its consultants to the main contractor (i.e. it assumes that any such information will have been provided by the client to the main contractor prior to conclusion of the contract). It is also worth noting that JCT 2005 MPCC does not envisage that the process of main contractor design submissions will be used to clarify cost issues, and assumes that at the date of the contract the main contractor will have had enough design information to provide a firm price. Only in the event of particular comments raised by the client on main contractor design submissions is it contemplated that a change (with cost and time consequences) might arise[790].

JCT 2005 Design and Build has incorporated a 'Contractor's Design Submission Procedure'[791] similar to that contained in JCT 2005 MPCC, but on a more restricted basis, as JCT 2005 Design and Build is intended to come into existence only on commencement of the construction phase of a project[792]. It is interesting to note that JCT 2005 Design and Build envisages submission of design documents pursuant to the Contractor's Design Submission Procedure as a precondition to commencement of particular work such that 'the Contractor shall not

---

[786] JCT 2005, SBC/Q, 2.2.
[787] JCT 2005 MPCC, clause 12.1.2.
[788] JCT 2005 MPCC, clause 15.1.
[789] JCT 2005 MPCC, Guidance Notes, Section 39.
[790] JCT 2005 MPCC, clauses 6.8 and 6.9.
[791] JCT 2005 Design and Build, clause 2.8 and Schedule 1.
[792] For example, JCT 2005 Design and Build includes Contract Particulars with a specific 'Date of Possession' and 'Date for Completion' and lacks any provisions or procedures by which either of these dates could be conditional upon preconstruction phase design activities.

commence any work to which such a document relates before that procedure has been complied with'[793]. However, to apply such procedure to the preconstruction phase of the project would require substantial amendment to JCT 2005 Design and Build in order to establish a two-stage appointment of the main contractor.

Another document in the JCT 2005 suite which provides for potential contractor design input is its Framework Agreement[794]. If entered into by a main contractor as a supplementary document to a JCT 2005 building contract, the JCT 2007 Framework Agreement encourages the contractor to offer value engineering contributions by way of changes to the relevant works 'which if implemented would result in financial benefits to the Employer'[795]. Approved value engineering proposals would result in a change being instructed under the relevant building contract and create the potential for shared financial benefits[796]. However, there is no preconstruction phase recognised in the JCT Framework Agreement, and the potential for the appointed contractor to contribute to value engineering appears to be restricted to ideas for changes that are put forward while the project is already being constructed on site. By this stage of a project the thinking time required for value engineering has substantially passed and any improvements offered are likely only to be marginal.

In addition, the JCT 2007 Framework Agreement offers no agreed financial incentives for value engineering, but only an 'agreement to agree' these after the parties have invested their efforts in coming forward with new ideas. Commercially, this does not appear a realistic expectation and it may in any event be unenforceable for lack of certainty[797].

JCT PCSA includes in its heads of Pre-Construction Services 'value engineering/buildability advice'[798] and envisages that the contractor 'will assist with final development of the design'[799]. It provides for the contractor to be paid a fee and agreed expenses for such services[800]. JCT PCSA(SP) also includes in its heads of Pre-Construction Services 'Specialist design development'[801]. Neither agreement contains detailed descriptions of design processes.

---

[793] JCT 2005 Design and Build 2005, clause 2.8.
[794] JCT 2007 Framework Agreement.
[795] JCT 2007 Framework Agreement, clause 17.1.
[796] JCT 2007 Framework Agreement, clause 17.1.
[797] See Chapter 2 Section 2.1 (The conditional preconstruction phase agreement) regarding contractual status and uncertainty.
[798] JCT PCSA, Annex B.
[799] JCT PCSA, Guidance Notes, 25.
[800] JCT PCSA, clause 6.
[801] JCT PCSA(SP), Annex B.

*Appendix B*

## 2.6 JCT CE design development

JCT CE does not contain provisions describing a preconstruction phase design development process. It includes general reference to provision of 'copies of all designs' and allowance of 'a reasonable time for comment'[802].

# 3 Two-stage pricing

## 3.1 GC/Works/1 two-stage pricing

Two-stage pricing is contemplated in GC/Works/1 Two Stage Design and Build, although this form does not contain detailed procedures. Significant features of the GC/Works process include:

- Appointment of the main contractor on a first-stage basis to develop designs in accordance with an agreed programme[803];
- The requirement that, 56 days in advance of the times shown in the programme for certification of completed designs and agreement of a contract sum, the main contractor is to submit 'a fully written itemised lump sum quotation of the proposed Contract Sum, based as far as possible on the rates and prices contained within the Pricing Document'[804];
- Agreement of the contract sum either by acceptance of the lump sum quotation or by measurement and valuation by the project quantity surveyor subject to the main contractor's right of objection[805];
- Sign-off of the agreed contract sum prior to the main contractor obtaining possession of the site[806].

## 3.2 NEC3 two-stage pricing

NEC3 includes a range of pricing options of which Option C (Target Contract With Activity Schedule) could fit particularly well with two-stage pricing[807]. However, NEC3 does not contain provisions describing a two-stage pricing process.

---

[802] JCT CE, clause 4.19.
[803] GC/Works/1 Two Stage Design and Build, clause 10B(1).
[804] GC/Works/1 Two Stage Design and Build, clause 10B(2).
[805] GC/Works/1 Two Stage Design and Build, clauses 10B(3), (4), (5), (6), (7), (8) and (9).
[806] GC/Works/1 Two Stage Design and Build, clause 34.
[807] NEC3 Option C: Target contract with activity schedule.

## 3.3 PPC2000 two-stage pricing

A two-stage pricing process is set out in PPC2000 in a manner which runs in parallel with the PPC2000 procedure for development of the supply chain. Key features of the PPC2000 two-stage pricing process are as follows:

- The main contractor is selected and appointed on the basis of an agreed budget for the project agreed profit and overheads and any other elements of the price that can be established at that stage[808];
- The main contractor then works with the client and consultants in developing designs and risk assessments sufficient for the main contractor to submit business cases or invite competitive prices from prospective subcontractors and suppliers, using enquiry documents and subcontracts that are approved by the client and with participation as agreed in the selection process by the client and other team members[809];
- Prices deriving from such exercises are included in the overall agreed price, with supporting details set out in the price framework[810];
- Pre-conditions to commencement of the project on site include finalisation of an agreed price supported by a price framework[811];
- The main contractor agrees to implement the construction phase of the project in consideration of the agreed price, subject only to such increases and decreases as may reflect incentivised savings or agreed changes or agreed events of delay and disruption[812].

## 3.4 Perform 21 two-stage pricing

Perform 21 does not contain provisions describing a two-stage pricing.

## 3.5 JCT 2005 two-stage pricing

JCT 2005 does not contain provisions describing a two-stage pricing process.

JCT PCSA includes in its heads of Pre-Construction Services 'Cost Advice'[813] and envisages that the contractor will assist the client with

---

[808] PPC2000 Partnering Terms, clauses 12.3 and 12.4.
[809] PPC2000 Partnering Terms, clauses 10.4, 10.5, 10.6 and 10.7.
[810] PPC2000 Partnering Terms, clauses 12.6 and 12.7.
[811] PPC2000 Partnering Terms, clause 14.1(vii).
[812] PPC2000 Partnering Terms, clauses 12.10, 15.2, 17 and 18.
[813] JCT PCSA, Annex B.

*Appendix B*

'specialist tender documents and with the arrangements necessary to obtain sub-contract tenders for the Contractor's second stage bid'[814]. It does not describe the specialist tender process or provide for pre-agreement of the contractor's profit and overheads in advance of its second stage bid, and it states that the client 'is under no obligation to accept any Second Stage Tender'[815]. This leaves considerable scope for misunderstandings and disagreements unless more detailed provisions are added and unless clearer links are created with the construction phase building contract. To obtain the benefit of a contractor's commitment to preconstruction phase services, it is necessary to provide a system whereby if the contractor performs the agreed services and establishes a price within an agreed budget, then it will be awarded a construction phase building contract.

## 3.6 JCT CE two-stage pricing

JCT CE does not contain provisions describing a two-stage pricing process.

# 4 Risk management

## 4.1 GC/Works/1 risk management

An amendment in 2000 to GC/Works/1 Two Stage Design and Build under the heading 'Risk Management', required the main contractor to have submitted with its tender a 'risk analysis' including a 'risk register' and 'financial analysis of the risks identified in the risk register', and to update both of these on a regular basis[816]. The reference to a document submitted with the main contractor's tender is historical and does not refer to any input by the client or any consultant. Nor does it recognise the creation or use of a mutually agreed risk register. The clause requires that the risk register and financial analysis be updated in reports prior to team meetings, again by the main contractor alone. It is difficult to see how the main contractor's compliance with this provision would itself contribute to joint risk management.

---

[814] JCT PCSA, Guidance Notes, 25.
[815] JCT PCSA, clause 2.7.2.
[816] GC/Works/1 (2000), Risk Management (amends Condition 1A (Fair dealing and teamworking)).

## 4.2 NEC3 risk management

NEC3 has introduced the requirement for a 'Risk Register' describing risks listed at the point of contract and those which either the project manager or the main contractor has notified as an early warning matter – including 'a description of the actions which are to be taken to avoid or reduce the risk'[817].

In place of the NEC2 'early warning meeting', NEC3 requires a 'risk reduction meeting' which can be initiated by the project manager or the main contractor and requires those who attend to cooperate in:

- 'Making and considering proposals for how the effect of the registered risks can be avoided or reduced;
- Seeking solutions that will bring advantage to all those who will be affected;
- Deciding on the actions which will be taken and who, in accordance with this contract; will take them;
- Deciding which risks have now been avoided or have passed and can be removed from the Risk Register'[818].

The provisions of NEC3 describe a joint risk management process. However, this remains confined to the duration of the relevant NEC3 contract, and the presumption remains (for example by virtue of a specified 'Completion Date' for the works[819]) that the NEC3 contract will be entered into to govern only the construction phase of the project, after the time when the primary opportunities exist for joint risk management.

Other points to note in the NEC3 approach to risk management include the fact that only the project manager and the contractor can instruct attendance at a risk reduction meeting, with no explicit role for the employer. This is logical as NEC3 envisages that the project manager's function is representative of the employer rather than as an independent party, but it emphasises the absence of the employer from active participation in the project. Also, the amendment of clause 16 (Early warning) in NEC3 to the effect that any meeting consequent on an early warning is a 'risk reduction meeting' rather than an 'early warning meeting' appears to restrict the previous availability of the NEC2 early warning system as a means to resolve differences between the parties rather than only to reduce risks[820].

---

[817] NEC3 core clause 11.2(14).
[818] NEC3 core clause 16.3.
[819] NEC3 core clause 30.1.
[820] NEC2 and NEC3, core clauses 16.2 and 16.3.

*Appendix B*

## 4.3 PPC2000 risk management

PPC2000 identifies risk management as a responsibility of team members and defines it as 'a structured approach to ensure that risks are identified at the inception of the Project, that their potential impacts are allowed for and that where possible such risks and their impacts are minimised'[821]. PPC2000 precludes any risk pricing until such time as the relevant risk has been reviewed by all team members with proposals for its elimination, reduction, insurance, sharing or apportionment and for removal or reduction of the relevant risk contingency[822]. It also provides for all project team members:

'To analyse and manage risks in the most effective ways including:

(1) Identifying risks and their likely costs;
(2) Eliminating or reducing risks and their costs;
(3) Insuring risks wherever affordable and appropriate;
(4) Sharing or apportioning risks according to which one or more Partnering Team members are most able to manage such risk'[823].

PPC2000 states a clear methodology for risk management in its annotated model risk register[824]. A copy of the PPC2000 form of risk register is set out in Appendix D. In addition, the PPC2000 Guide states that:

'The analysis and management of risks relevant to the Project should be by a methodology agreed by the Partnering Team prior to signing the Project Partnering Agreement and reflected in activities described in the Partnering Documents, for example the preparation and agreement of a risk register with an agreed action plan as to how Partnering Team members will deal with the risks identified and any prospective risk contingencies'[825].

## 4.4 Perform 21 risk management

Perform 21 does not contain provisions describing a risk management process.

---

[821] PPC2000, Appendix 1.
[822] PPC2000 Partnering Terms, clause 12.9.
[823] PPC2000 Partnering Terms, clause 18.1.
[824] PPC2000, Appendix 7.
[825] Mosey (2003), 39.

## 4.5 JCT 2005 risk management

The JCT 2007 Framework Agreement provides for 'collaborative risk analysis' to be undertaken in respect of each project if so required by a client 'Enquiry' or if otherwise agreed[826]. Relevant features of the JCT Framework Agreement's approach to risk analysis and risk allocation are as follows:

- It expects identification of 'significant potential risks' affecting cost, programme or quality, determination of 'the likelihood of such risks occurring' and 'the seriousness of the likely consequences thereof' and who is 'best able to manage such risks'[827].
- It then envisages the client drawing up 'a risk allocation schedule or matrix'[828].
- It provides for periodic review of the risk allocation schedule or matrix during the life of each project. However, given the subsidiary nature of the Framework Agreement to the Underlying Contracts, there will need to be an additional mechanism in the Underlying Contracts to accommodate any different approach to risk management that comes out of such regular reviews[829]. The previous JCT 2005 Framework Agreement provided that there was scope for altering risk allocation, remuneration and programme when concluding Underlying Contracts. These provisions have been removed from the JCT 2007 Framework Agreement.

The risk management provisions in the JCT 2005 Framework Agreement, since watered down in the 2007 version, represented the first JCT contractual system governing preconstruction phase risk management activities. It is particularly interesting that the 2005 provisions recognised that the parties could, as a result of such activities, agree the different distribution of risk to that set out in a published JCT 2005 building contract.

The principal weakness in the JCT 2007 Framework Agreement risk management proposals is now the unanswered question of how the risk allocation and management strategy for a project will affect the main contractor's price and programme, and what happens if the parties do not agree. Although the JCT 2007 Framework Agreement clearly contemplates a working relationship in relation to risk management in advance of Underlying Contracts, it lacks clarity as to the means by which the terms of those Underlying Contracts will be finalised and put in place.

---

[826] JCT 2007 Framework Agreement, clause 14.
[827] JCT 2007 Framework Agreement, clause 14.1.
[828] JCT 2007 Framework Agreement, clause 14.2.
[829] JCT 2007 Framework Agreement, clauses 14.4 and 6.

*Appendix B*

JCT PCSA makes no reference to the contractor's involvement in joint risk management. This is a serious omission as the absence of joint risk management will be more likely to leave the client's and the contractor's differing risk assumptions unaffected by the early contractor appointment, with priced risk contingencies consequently more likely to remain embedded in second-stage tender prices.

## 4.6 JCT CE risk management

Although JCT CE provides for a risk register and risk allocation schedule[830], the opportunities for joint risk management under the JCT CE are limited. Unless a separate JCT CE preconstruction phase agreement is put in place, there will be no contractual commitment to undertake preconstruction risk management activities, no timetable for such activities and no link to related design development and pricing activities.

# 5 *Communications*

## 5.1 GC/Works/1 communications

GC/Works/1 Two Stage Design and Build does not contain provisions describing communication systems by way of early warning, use of a core group or otherwise.

## 5.2 NEC3 communications

The contractual duty to warn was first set out in earlier editions of NEC3, and is now addressed as follows:

'The Contractor and the Project Manager give an early warning by notifying the other as soon as either becomes aware of any matter which could:

- Increase the total of the Prices;
- Delay completion;
- Delay meeting a Key Date; or
- Impair the performance of the works in use'[831].

---

[830] JCT CE Sections 5.1, 5.2, 5.3 and 5.4.
[831] NEC3, core clause 16.1.

An additional duty of warning is expressed in slightly different wording in NEC3 Option X12 whereby 'Each Partner gives an early warning to the other Partners when he becomes aware of any matter that could affect the achievement of another Partner's objectives stated in the Schedule of Partners'[832].

Although NEC3 provides for operation of a 'Core Group', under NEC3 Option X12, receipt and consideration of early warning is not one of its stated functions[833]. NEC3 Option X12 states that 'The Core Group comprises the Partners listed in the Schedule of Core Group Members', although the reference to 'Partners' creates some ambiguity as to whether the core group comprises individuals or organisations[834]. If individuals are not identified as core group members, then the NEC3 Core Group would not assist communication in the ways envisaged in Chapter 5, Section 5.3.3 (Creation of a contractual core group).

As to the functions of the core group, NEC3 Option X12 states that 'The Core Group acts and takes decisions on behalf of the Partners on those matters stated in the Partnering Information'[835]. NEC3 does not include model Partnering Information, and detailed attention will be required in drafting this Partnering Information so as to establish terms of reference for the Core Group that are integrated with the remainder of the NEC3 contracts. For example, it could allocate to the Core Group all or any of the equivalent functions described in PPC2000, for example as to approval of proposals or review of early warning notices.

NEC3 Option X12, states the following functions of the Core Group:

- 'The Core Group decides how they will work and decides the dates when each member joins and leaves the Core Group'. There is no timescale for this, so there is the risk that early Core Group meetings will need to devote significant time to working out their own terms of reference and procedures. This lack of clarity also creates uncertainty for any team members who are not represented on the Core Group.
- 'The Core Group may give an instruction to the Partners to change the Partnering Information'. This creates further scope for confusion regarding the functions of the Core Group and their influence over team members who are not represented on it.
- 'The Core Group prepares and maintains a timetable showing the proposed timing of the contributions of the Partners', revises it as required and 'The Contractor changes his programme if it is necessary to do so in order to comply with the revised timetable.' This is

---

[832] NEC3 Option X12, clause X12.3(3).
[833] NEC3 Option X12, clauses X12.1, X12.2 and X12.3.
[834] NEC3 Option X12, clause X12.1(3).
[835] NEC Option X12, clause X12.2(3).

*Appendix B*

an important function as regards programming of the project and is discussed further below.
- 'A Partner gives advice, information and opinion to the Core Group and to other Partners when asked to do so by the Core Group.' This could prove onerous in terms of the resources required for a partner to give the required advice, information and opinion, and the clause is not linked to the NEC3 change procedure or any other means to allow the cost of such resources to be recovered.
- 'A Partner notifies the Core Group before sub-contracting any work'. This does not create a right of the Core Group to approve the sub-contracting proposal, and appears to be for information only[836].

The stated power of the Core Group to instruct a change in the main contractor's programme so as to comply with a revised timetable appears to be a direct challenge to the authority of the project manager under NEC3 to deal with all change procedures. If a proposed change to the main contractor's programme resulting from the Core Group's timetable (or a revision to it) gives rise to the main contractor quoting increased prices rather than the reduced prices anticipated, it is assumed that under NEC3 the project manager would retain its power to withdraw the proposed change. NEC3 Option X12, however, does not recognise this power and specifically requires the main contractor to make the necessary changes to its programme.

This illustrates another omission in the NEC3 Core Group provisions, namely a clear understanding as to how they reach their decisions. NEC3 Option X12 does not state how many core group members need to attend a meeting in order to constitute a quorum, nor does it state how many need to vote in favour of a proposal in order for the Core Group to reach a decision. These matters are left for the Core Group to decide with no time limit for making these decisions. This creates the following problems:

- If there is no provision as to the quorum or majority required for the Core Group to decide how it will work and if, as a consequence, these matters are not dealt with prior to creation of the relevant NEC3 contracts, then in the absence of consensus on these issues the Core Group will not be able to get started.
- If the Core Group does not adopt a decision-making procedure based on unanimous agreement, then it could reach decisions by majority vote which have an adverse effect on a dissenting team member, for example on the main contractor as regards a proposed change to its programme.

---

[836] NEC3 Option X12, clauses X12.2 and X12.3.

- If the Core Group does not provide for its decisions only to be reached by those members attending a meeting, then all its decisions could be blocked by non-attendance of any Core Group member.

## 5.3 PPC2000 communications

A contractual duty to warn is set out in PPC2000 under which:

- Team members are required to warn each other and the project manager of 'any error, omission or discrepancy of which they become aware between the Partnering Documents'[837].
- Team members are also required to warn each other as soon as they are 'aware of any matter adversely affecting or threatening the Project or that Partnering Team member's performance under the Partnering Contract'[838].

Under PPC2000 the party submitting the warning is required to put forward proposals (within the scope of its agreed role, expertise and responsibilities) for dealing with the problem, and the project manager (itself a party to PPC2000) convenes a meeting of the Core Group unless an appropriate course of action can be agreed without a meeting[839].

PPC2000 requires the team members to establish the Core Group 'who shall meet regularly to review and stimulate the progress of the Project and the implementation of the Partnering Contract'[840] as well as to fulfil specific agreed functions stated in the contract.

Features of the PPC2000 Core Group include:

- A system whereby the project manager convenes meetings at not less than five working days notice (unless all Core Group members agree a shorter period) with a stated agenda;
- Provision that all team members who are signatories to PPC2000 are entitled to attend a Core Group meeting (even if not represented on the Core Group itself) and that all team members are required to comply with any decision of the Core Group made within the scope of its agreed functions;

---

[837] PPC2000 Partnering Terms, clause 2.5.
[838] PPC2000 Partnering Terms, clause 3.7.
[839] PPC2000 Partnering Terms, clause 3.7. PPC2000 lists the named members of the Core Group and restricts change in Core Group membership without all team members' consent.
[840] PPC2000 Partnering Terms, clause 3.3.

*Appendix B*

- Decisions of the Core Group are to be by 'Consensus' of all Core Group members present at a meeting (Consensus being defined as 'unanimous agreement following reasoned discussion');
- Team members are obliged to ensure that those of their employees who are Core Group members attend its meetings and fulfil their agreed functions[841].

As to the role of the PPC2000 Core Group in relation to preconstruction phase processes:

- The Core Group has a role alongside the client in approving each stage of design development, including proposals for value engineering[842].
- The Core Group has a role in considering, with the client, business cases put forward for single source procurement by the main contractor whether as proposed direct labour packages or as preferred subcontractors or suppliers, and in selecting subcontractors and suppliers who offer best value by way of subcontract tenders[843].
- The Core Group has a role in investigating the potential for cost savings and added value, including through risk management, and the agreement of any incentives considered appropriate for this purpose, including shared savings, shared added value or shared pain/gain[844].

In addition to the joint preconstruction phase processes, the Core Group under PPC2000 is also the recipient of early warning notices and related proposals dealing with the problems notified[845].

## 5.4 Perform 21 communications

The Perform 21 Partnering Agreement contains an early warning system by which 'Each Partner will give an early notice to the other Partners when aware of any matter which may cause delay or impair performance in the achievement of the objectives set out in the Partnering Charter'[846]. The Partnering Charter remains to be prepared by the parties as a bespoke document, and clear definition of objectives in that Partnering Charter will be necessary if the parties are going to

---

[841] PPC2000 Partnering Terms, clauses 3.3, 3.4, 3.5 and 3.6 and Appendix 1.
[842] PPC2000 Partnering Terms, clauses 8.3 and 8.8.
[843] PPC2000 Partnering Terms, clauses 10.4, 10.5 and 10.6.
[844] PPC2000 Partnering Terms, clauses 12.10, 13.1 and 13.2.
[845] PPC2000 Partnering Terms, clauses 2.5 and 3.7.
[846] Perform 21 PSPCP, clause 5.

be able to establish when an early warning is appropriate. In addition, the early warning obligation is not linked to a series of consequent actions, whether by way of a meeting to agree a solution or otherwise. Accordingly, it is unlikely that a party will give early warning if this could in any way adversely affect its commercial position, as the response to such early warning could simply be that the recipient uses the information to support a claim or dispute.

A Core Group has been recognised in the Perform 21 Partnering Agreement[847]. This agreement provides that 'The Core Group will act for the benefit of the Project and with this aim will take the high level decisions on behalf of the Partners, as set out in the Decision Making Process in Appendix Part 4'[848]. However, Appendix Part 4 is left blank, and there is no reference to the status of the Core Group in any of the related building contracts and professional appointments that constitute the remainder of the Perform 21 suite of contracts. Accordingly, considerable work will be required in drafting terms of reference for the Core Group to deal with the matters described above, and also in creating consistent provisions in the related building contracts and professional appointments to reflect the authority of the Core Group and the effect of the decisions it makes.

## 5.5 JCT 2005 communications

The JCT 2005 suite does not include in its building contracts any provisions describing communication systems by way of early warning, use of a core group or otherwise.

However, the JCT 2007 Framework Agreement introduced the requirement for each party to provide the other with 'a detailed organisation and management diagram setting out and explaining their own internal organisational and management structures in detail, including particular of the roles, responsibilities and limits of authority of all key management personnel'[849]. The JCT 2007 Framework Agreement also requires the parties to establish links between this diagram and the individuals actually exercising authority under any Underlying Contract, revising the diagram as necessary, and that whenever decisions need to be made concerning projects governed by those Underlying Contracts 'the Parties will endeavour to ensure that those responsible, with authority to make such decisions, are fully briefed and on hand to make such decisions as appropriate'[850]. All of this

---

[847] Perform 21 PSPCP, clause 7.
[848] Perform 21 PSPCP, clause 8.
[849] JCT 2007 Framework Agreement, clause 8.2.
[850] JCT 2007 Framework Agreement, clause 8.5.

makes good sense, although it is likely to be insufficient for parties only to 'endeavour' to ensure that responsible individuals making key decisions are fully briefed and available. As those individuals will be the employees or paid consultants of the relevant party, outright obligations would appear to be more appropriate.

The JCT 2007 Framework Agreement requires the parties to establish a 'common communications protocol' to promote 'clear and effective communication and the dissemination and ready availability of information essential to the success of each of the Tasks'[851] governed by the separate Underlying Contracts. However, it qualifies creation of the communications protocol as 'Without in any way detracting from or affecting the specific notice and communication requirements of the Underlying Contracts'[852]. Unless the Underlying Contracts include their own communications protocol (and there is no reference to this in the remainder of the JCT 2005 suite of contracts), there is potential for considerable confusion as to what, if any, authority derives from the JCT 2007 Framework Agreement communications protocol, and it is therefore unlikely that the parties will invest the necessary effort in creating and maintaining it.

The JCT 2007 Framework Agreement also includes the unusual requirement that 'the Parties will at all times endeavour to keep things factual and to the point and will avoid self-serving statements, assertions of blame and/or emotive or provocative language'[853]. This represents a significant departure for the JCT as it is clearly an attempt to use a contract document to create a collaborative culture for a project rather than simply to deal with legal roles and responsibilities. It is unlikely that such vague wording could be construed as a legal obligation even if it was not qualified by the word 'endeavour' and even if it was also included in the Underlying Contracts[854].

The delegation of authority under the 'detailed organisation and management diagram' and the application of the 'communications protocol' need to be legally binding if the team are to rely on them in the running of a project. However, this is apparently not the JCT's intention. Instead, the qualification created by the word 'endeavour' plus the provision that the JCT 2007 Framework Agreement will always be subsidiary to the Underlying Contracts[855], dilute and confuse the

---

[851] JCT 2007 Framework Agreement, clause 12.1.
[852] JCT 2007 Framework Agreement, clause 12.1.
[853] JCT 2007 Framework Agreement, clause 12.2.
[854] It does not appear in the JCT 2005 suite or in JCT CE. See also Chapter 2, Section 2.1 (The conditional preconstruction phase agreement) regarding enforceability and uncertainty.
[855] JCT 2007 Framework Agreement, clause 6.

enforceability of communication provisions that are of central importance to the project.

The early warning provision set out in the JCT 2007 Framework Agreement requires that 'each of the Parties will promptly warn the other Party in writing of any matter or concern of which he becomes aware which in that Party's reasonable opinion is likely to affect the out-turn cost or programme or the quality or performance of any Tasks'[856]. This duty is expressly stated not to detract from or affect the notice requirements of Underlying Contracts. More significantly it does not state what will be done when an early warning notice is received. There is no provision for use of a core group under the JCT 2007 Framework Agreement or any of the JCT 2005 contracts, nor is there any reference in the JCT 2005 contracts to early warning or the actions that should be taken if such early warning is given.

As a consequence, it is unlikely that, for example, a main contractor suffering a problem that is causing delay will give early warning of that problem as required by the JCT 2007 Framework Agreement, if the only apparent contractual consequence of that warning in its Underlying Contract will be to undermine any related contractor claim for extension of time and loss and expense, and to give rise instead to a liability for liquidated and ascertained damages.

JCT PCSA includes in its heads of Pre-Construction Services 'Establishment of management and communication systems for the Construction Phase (including external links)'[857]. However, it does not put these systems in place to govern or support performance of the Pre-Construction Services themselves or state how such systems will be reconciled with the terms of the construction phase building contract. JCT PCSA does, however, include a duty on the part of the contractor to warn of any 'inconsistency or divergence' in relevant documents or any 'delay or impediment in performing the Pre-Construction Services'[858]. There is no duty to warn on the part of the client.

## 5.6 JCT CE communications

JCT CE does not contain any provisions describing communication systems by way of early warning, use of a core group or otherwise.

---

[856] JCT 2007 Framework Agreement, clause 19.
[857] JCT PCSA, Annex B.
[858] JCT PCSA, clause 2.3.3.

*Appendix B*

# 6 Programmes

## 6.1 GC/Works/1 programmes

GC/Works/1 Two Stage Design and Build requires that 'the Programme shows the Design Process Event and the time for its certification' as well as the subsequent construction phase activities[859]. As this programme governs both the preconstruction phase and the construction phase, it will need to be carefully structured to take account of possible delays in commencement of the construction phase while the client approves the main contractor's design documents.

GC/Works/1 Two Stage Design and Build makes clear that the agreed programme is to be adhered to by the main contractor both in relation to the preconstruction phase activities and the construction phase activities, which comprise 'the sequence in which the Contractor proposes to execute the Works, details of any temporary work, method of work, labour and plant proposed to be employed, and events, which, in his opinion, are critical to the satisfactory completion of the Works' and which should include among other things 'reasonable periods of time for the provision of information required from the Employer'[860]. Accordingly, the contractually binding GC/Works/1 programme is more detailed than the JCT Information Release Schedule or the PPC2000 Project Timetable, and could bind the client to the main contractor's detailed methods of working in a way that could render the client vulnerable to claims equivalent to those described in the Yorkshire Water Authority case[861].

## 6.2 NEC3 programmes

Preconstruction phase processes under NEC3 would need to be governed by an NEC3 Professional Services Contract. This requires a 'programme', but without specifying particular activities, and allows the option for the programme to be provided after work has commenced[862]. Detail would need to be added to describe relevant preconstruction phase processes and to clarify and integrate the contractual effect of particular deadlines under the different NEC3 Professional Services Contracts entered into with the main contractor and with each consultant.

---

[859] GC/Works Two-stage Design & Build, clause 33(1).
[860] GC/Works/1 Two-stage Design & Build, clause 33(1).
[861] *Yorkshire Water Authority v. Sir Alfred McAlpine & Son (Northern) Ltd* (1985), 32 BLR 115. See also Chapter 5, Section 5.4.5 (Early agreement of construction phase programmes).
[862] NEC3 Professional Services Contract, core clause 31.1.

NEC3 requires a 'programme'[863], but only in relation to the construction phase and again with the option that this can be provided after the construction phase has commenced. This potential time lag between entering into a contractual commitment to build a project and generating a programme for approval by the client and project manager increases the likelihood of the parties disagreeing over key timing issues and of the programme never in fact being finalised.

Frances Forward takes the view that NEC3 'makes the programme an integral contract document which is kept up to date and can be used as a project management tool, both for monitoring progress accurately and in relation to incentivisation'[864]. However, the efficacy of the NEC3 programme as a project management tool will be much reduced if it is not agreed and implemented from the commencement of the project.

An additional requirement for a programme appears in NEC3 Option X12 by which the Core Group prepare and maintain a 'timetable' that shows the proposed timing of the contributions of each team member, and issue a copy of this timetable to team members each time it is revised[865]. Significantly, the NEC3 Option X12 provision for the main contractor to change its programme if necessary in order to comply with the revised timetable prepared by the Core Group is on the basis that 'Each such change is a compensation event which may lead to reduced Prices'[866].

NEC3 (in contrast to NEC2) introduces in its building contract and in the NEC3 Professional Services Contract the concept of a 'Key Date' which is defined as 'the date by which work is to meet the Condition stated'[867]. NEC3 requires the main contractor to undertake the work 'so that the Condition stated for each Key Date is met by the Key Date'[868]. However, Key Dates are not separated from the programme but instead form an integral part of the programme – and thus, if not agreed in advance of the NEC3 contract, would be subject to later agreement (or not) during the period following commencement of the project on site[869].

## 6.3 PPC2000 programmes

PPC2000 provides for a preconstruction phase programme known as the 'Partnering Timetable'[870] signed with the Project Partnering

---

[863] NEC3 core clause 31.1.
[864] Forward (2002), 6.
[865] NEC3 Option X12, clause X12.3(7).
[866] NEC3 Option X12, clause X12.3(7).
[867] NEC3, core clause 11.2(9).
[868] NEC3, core clause 30.3.
[869] NEC3, core clause 31.
[870] PPC2000 Partnering Terms, clause 6.1.

*Appendix B*

Agreement by the client and the main contractor together with all consultants and any previously selected subcontractors and suppliers who are members of the 'Partnering Team'. A copy of the model PPC2000 form of Partnering Timetable is set out in Appendix E. The Partnering Timetable is defined as 'governing the activities of the Partnering Team members in relation to the Project prior to the date of the Commencement Agreement'[871] (i.e. the document confirming that the project is ready to start on site) and PPC2000 creates a contractual obligation on all partnering team members 'to undertake their agreed activities in relation to the Project ... regularly and diligently in accordance with the Partnering Timetable'[872].

PPC2000 provides for a construction phase programme known as the 'Project Timetable'[873] to be produced by the main contractor and agreed by all other Partnering Team members as a precondition to commencement of the project on site. As in the case of the Partnering Timetable, the Project Timetable will be contractually binding on the client and the main contractor and those consultants and subcontractors and suppliers who are signatories to PPC2000, subject to:

- Any agreed pre-conditions to its implementation;
- Any agreed arrangements for acceleration or postponement;
- The agreed contractual change procedure;
- The agreed contractual procedure in respect of events of delay and disruption;
- Statutory rights of suspension for non-payment and agreed rights of suspension or abandonment of the project[874].

PPC2000 distinguishes the Project Timetable from 'supporting method statements and procedures' that are to be submitted separately by the main contractor, but which do not have contractual status[875].

## 6.4 Perform 21 programmes

The Perform 21 Partnering Agreement and building contracts do not contain provisions for the creation of contractually binding programmes or key date schedules in respect of preconstruction phase or construction phase activities, although the Prestart Agreement

---

[871] PPC2000, Appendix 1.
[872] PPC2000 Partnering Terms, clause 6.1.
[873] PPC2000 Partnering Terms, clause 6.2.
[874] PPC2000 Partnering Terms, clauses 6.6, 14, 17, 18, 20.17 and 26.6.
[875] PPC2000 Partnering Terms, clause 6.2.

provides for a programme to be prepared by the contractor and submitted for acceptance by the client[876].

## 6.5 JCT 2005 programmes

The JCT 2005 contracts do not provide for any preconstruction phase construction programme or key dates schedule as they do not deal with preconstruction phase activities.

The JCT 2005 contracts provide for the main contractor to submit a 'master programme' as soon as possible after execution of the contract, but without affecting the parties' contractual obligations[877]. The JCT 2005 contracts provide for agreement of a 'Date of Possession' and a 'Date for Completion'[878], but all other interaction between the parties as described in the main contractor's master programme is free from any time limit – as, for that matter, is the main contractor's obligation to create the master programme in the first place.

The JCT 2005 contracts do, however, contain an additional programming document known as the 'Information Release Schedule' which can obligate parties to release specific information to each other at specific times[879]. If an Information Release Schedule is used and is suitably populated, it can deal with the timely release to the main contractor of design details not available when the contract was created, including those necessary for the main contractor to firm up provisional sums. It can also provide deadlines for the submission by the main contractor of prices in respect of such provisional sum items in sufficient time for the client to go through appropriate approval processes before expenditure is committed.

In the absence of an Information Release Schedule, the JCT 2005 contracts propose only a general requirement for further drawings or details to be released to the main contractor (including those necessary for expenditure of provisional sums) 'at the time it is reasonably necessary for the Contractor to receive them, having regard to the progress of the Works'[880]. This leaves considerable scope for disagreement as to what is 'reasonable'.

A further reference to time limits during the construction phase under the JCT 2005 contracts appears in the 'Contractor's Design Submission Procedure' which requires the main contractor to submit required designs 'in sufficient time to allow any comments of the

---

[876] Perform 21 PSPC 10, clause 2.1.
[877] JCT 2005 SBC/Q, clause 2.9.1.
[878] JCT 2005 SBC/Q Contract Particulars.
[879] JCT 2005 SBC/Q, clause 2.11.
[880] JCT 2005 SBC/Q, clause 2.12.

*Appendix B*

Architect/Contract Administrator to be incorporated'[881] prior to them being used for procurement or works. Although a 'sufficient time' is not specified, this could be set out in the Information Release Schedule, and the Contractor's Design Submission Procedure goes on to state clearly defined timescales for the approval procedures relating to Contractor's Design Documents as submitted.

The JCT MPCC provides for a 'design programme'[882] but otherwise assumes that implementation of the project will be within the main contractor's control and therefore provides only for an agreed 'Completion Date'[883].

All the provisions of the JCT 2005 contracts relate only to the mutual obligations of the client and main contractor and not to the related obligations of consultants. As JCT does not publish consultant appointments, it is necessary to review the compatibility of the JCT 2005 programming provisions with equivalent provisions set out in other standard forms of appointment such as those published by the RIBA and ACE[884]. Neither of these forms require consultants to perform their services in accordance with a programme, nor do they cross-refer to the JCT Information Release Schedule.

JCT PCSA provides for Pre-Construction Services to be performed in accordance with an agreed 'Programme'[885]. It also includes in its heads of Pre-Construction Services 'Programme preparation'[886]. However, this is misleading if intended to relate to preparation of the construction phase programme as it uses the defined term for the pre-construction 'Programme' which is listed as a document already in existence. JCT PCSA undermines the binding effect (and for this purpose the completeness) of its own programme by allowing the contractor to inform other team members 'in due time' of any information it requires 'that is not provided for in the Programme'[887].

## 6.6  JCT CE programmes

JCT CE does not contain provisions for the creation of contractually binding programmes or key dates schedules in respect of preconstruction phase or construction phase activities. It provides for a 'Project

---

[881] JCT 2005 SBC/Q, clause 2.9.3 and Schedule 1 Contractor's Design Submission Procedure.
[882] JCT MPCC, clause 12.3.
[883] JCT MPCC Project Particulars.
[884] RIBA (2004) and ACE (2002).
[885] JCT PCSA, clause 2.1.
[886] JCT PCSA, Annex B.
[887] JCT PCSA, clause 2.3.2.

Programme' but envisages that this may be prepared after creation of the contract[888].

## 7 Team integration

### 7.1 GC/Works/1 team integration

GC/Works/1 Two Stage Design and Build does not provide for any contractual links between the consultants, main contractor, subcontractors and suppliers under their respective two-party contracts, although it does comprise an integrated set of consultant appointments, building contracts and subcontracts.

### 7.2 NEC3 team integration

NEC3 is not a multi-party contract, but comprises an integrated set of consultant appointments, building contracts and subcontracts that can be more closely linked through the use of NEC3 Option X12 setting out agreed objectives of all parties and agreed joint working methods and incentivisation. Relevant features of NEC3 include commitment to 'common information systems'[889].

### 7.3 PPC2000 team integration

PPC2000 is a multi-party form of contract, creating a single contractual relationship to integrate commitments of all team members as defined by their respective roles, expertise and responsibilities as stated in the contract documents[890]. Relevant features of PPC2000 include:

- The contractual facility for referring issues to a single 'Partnering Adviser' accountable to all team members for 'provision of fair and constructive advice as to the partnering process, the development of the partnering relationships and the operation of the Partnering Contract'[891];
- Availability of alternative dispute resolution systems controlled by the team members themselves including a 'Problem-Solving Hierarchy' of named individuals at increasing levels of seniority,

---

[888] JCT CE, clause 4.19.
[889] NEC3 Option X12, clause X12.3(4).
[890] PPC2000 Partnering Terms, clause 1.3.
[891] PPC2000 Partnering Terms, clause 5.6(iv).

*Appendix B*

reference to the Core Group and reference to conciliation, mediation or any other form of alternative dispute resolution recommended by the Partnering Adviser[892].

## 7.4 Perform 21 team integration

The Perform 21 Partnering Agreement establishes mutual recognition of partnering values, subject to the team agreeing a bespoke 'Decision Making Process', and also establishes the following arrangements:

- Commitment to common information systems 'as far as reasonable';
- Agreement to comply with procedures to be included in a 'Procedures Document', which is another blank form requiring bespoke drafting and which (as it is not mentioned in the remainder of the suite of contracts) will also need to be reflected in the drafting of corresponding provisions in each consultant appointment, building contract and subcontract if it is to govern the way the project is implemented[893].

The collaborative principles set out in the Perform 21 Partnering Agreement are in some respects at odds with the traditional terms of the other Perform 21 Contracts. For example, the consultant appointment requires an outright transfer of copyright by the consultant to the client[894] rather than a licence, and is likely to be seen as favouring the client at the expense of what is commercially reasonable. Also, the Perform 21 building contracts allow for rights of termination only on the part of the client and not under any circumstances on the part of the contractor[895]. These one-sided provisions may make it hard for the parties to adopt and sustain the values expressed in the multi-party Partnering Agreement.

## 7.5 JCT 2005 team integration

JCT 2005 did not include a form of consultant appointment and therefore could not be used to establish contractual or other links between the consultants and the main contractor or its subcontractors and suppliers under their respective two-party contracts. JCT CA was

---

[892] PPC2000 Partnering Terms, clauses 27.2, 27.3 and 27.4.
[893] Perform 21 PSPCP, clauses 3, 8 and 9.
[894] Perform 21 PSPC 9, clause 14.0.
[895] For example, Perform 21 PSPC3, clause 32.

introduced at the end of 2008 as the first consultant appointment in the JCT suite. It includes provision for the consultant to 'liaise and cooperate fully with the other members of the Project Team'[896], which include other consultants, the contractor (or prospective contractor) and any specialists nominated by the client or lead consultant.

JCT PCSA also requires the contractor to 'liaise and cooperate fully with other members of the Project Team'[897].

## 7.6 JCT CE team integration

JCT CE's 'overriding principle' should influence the parties to each two-party contract, but it does not establish direct contractual links between the parties beyond their respective two-party contracts or set out any specific collaborative mechanisms.

JCT CE does, however, provide for the preparation by the project team of a 'project protocol' which is intended to set out the aims and objectives of the project team 'with regard to the delivery of the Project and the development of their working relationships'[898].

Confusingly, JCT CE also states that 'The provisions of any project protocol shall not create any contractual obligation and any failure to adhere to its terms shall not itself constitute a breach of this Contract'[899].

JCT CE includes an additional optional multi-party Project Team Agreement, but this provides that it will not create any mutual duties of care or liabilities between team members except in respect of payment pursuant to its risk and reward sharing arrangements[900].

---

[896] JCT CA, clause 2.3.
[897] JCT PCSA, clause 2.3.
[898] JCT CE, clause 2.6.
[899] JCT CE, clause 2.8.
[900] JCT CE Project Team Agreement, clause 2.9.

# APPENDIX C
# PRECONSTRUCTION PHASE PROCESSES UNDER STANDARD FORM FRAMEWORK AGREEMENTS

## Introduction

This appendix reviews the structure and relevant provisions of the two published standard form framework agreements, in order to consider the extent to which they describe preconstruction phase processes and operate as conditional preconstruction phase agreements for individual projects within the scope of the framework.

## JCT framework agreement

The publication of the JCT 2005 Framework Agreement as part of the JCT 2005 suite represented an intriguing new departure for the JCT. In its introduction to the new form, the JCT made reference to the 1998 report of the Construction Task Force *Rethinking Construction*[901] and cited its criticisms of:

(1) Procurement systems based on single projects with 'little opportunity for achieving incremental improvements in efficiency and/or effectiveness'; and
(2) Fragmentation of production 'roles and processes' in a way that inhibits 'effective and efficient teamworking, sharing of information and know-how'[902].

The JCT state that framework agreements can enable project participants to take a longer term view, to build and develop relationships, invest in products and processes and enhance commercial opportunities[903]. These are worthwhile goals[904]. To what extent did the JCT 2005 Framework Agreement, and its replacement, the JCT 2007 Framework Agreement, achieve them?

---

[901] Egan (1998).
[902] JCT 2005 Framework Agreement, Guide 1.

## A framework agreement or a partnering agreement?

The JCT 2005 Framework Agreement was not a framework agreement at all in the sense envisaged by the Public Contracts Regulations. Although clearly intended to span a number of different projects, it did not include in its 'Framework Particulars' a description of any particular type of project or programme of work awarded by the 'Employer' to the 'Service Provider'. Instead, it stated simply a 'Framework Start Date', and a 'Framework End Date'[905].

Further, the JCT 2005 Framework Agreement did not include any procedures for the creation of what it termed 'Underlying Contracts' governing specific projects, whether by competition or otherwise. Instead, it assumed that the machinery by which a project is identified, by which the Service Provider is selected for that project and by which the Underlying Contracts for each project put in place would all be dealt with by some other unstated means.

By contrast, the JCT 2007 Framework Agreement describes a system of 'Enquiry' in respect of the ordering of a series of 'Tasks' and is clearly intended to govern the award of a flow of work under successive Underlying Contracts[906].

However, consultants and contractors looking for long-term client commitments to a stable flow of work will not find it in the JCT 2007 Framework Agreement – due to the ability of either party to terminate at not less than one month's notice at any time[907].

The JCT 2005 Framework Agreement would be better described as a JCT partnering agreement, as its provisions comprised references to the working practices that have come to be associated with partnered projects. These included organisational structures and decision making, collaborative working, supply chain consultation, sharing of information and know-how, risk assessment and risk allocation, value engineering, change control, early warning and a team approach to problem solving/dispute resolution. Most of these provisions have been retained in the JCT 2007 Framework Agreement[908].

---

[903] JCT 2005 Framework Agreement, Guide 1.
[904] See also Chapter 7 (Increased preconstruction commitments under framework agreements).
[905] JCT 2005 Framework Agreement, clause 1 and Framework Particulars.
[906] JCT 2007 Framework Agreement, clauses 3 and 4. See also Chapter 7, Section 7.4 (Published forms of framework agreement).
[907] JCT 2007 Framework Agreement, clause 22.2.
[908] See also Appendix D.

*Appendix C*

## Binding or non-binding?

The JCT 2005 Framework Agreement was published in binding and non-binding versions. In providing a non-binding option the JCT risked confusing some clients, contractors and consultants as to whether such a document was appropriate to constitute a legal agreement. If not, then its provisions were optional, which would leave the parties uncertain as to whether they could rely on each other to adopt the good practices that the JCT 2005 Framework Agreement encouraged. This ambiguity was unnecessary and the non-binding option disappeared in the JCT 2007 Framework Agreement.

However, the JCT 2007 Framework Agreement continues to provide that it is subsidiary to the Underlying Contracts, which will prevail in the event of 'conflicting/discrepant provisions'[909]. Accordingly, the parties will be 'excused compliance' with the relevant provisions of the JCT 2007 Framework Agreement if, as is often the case, it contradicts their JCT 2005 building contract[910].

## Supply chain

Under the JCT 2007 Framework Agreement, the 'Provider' is required to 'endeavour' to achieve closer involvement by members of its 'Supply Chain' in matters such as 'design development', 'project planning', 'risk assessment' and 'value engineering'[911]. This looks promising, but the JCT 2007 Framework Agreement does not provide a timeframe or set of procedures within which any of these activities will occur. Specifically, it does not provide for early involvement of supply chain members during the preconstruction phase of a project when all of these activities can best be undertaken. In addition, it does not offer the contractor or the members of its supply chain any reward in return for providing this added value. Since a JCT Underlying Contract is put in place only when a project commences on site, there is no explanation in the JCT 2007 Framework Agreement as to how either the contractor or its supply chain members will be appointed on a specific project at a time when their added value could best be provided, namely during the preconstruction phase.

---

[909] JCT 2007 Framework Agreement, clause 6.
[910] JCT 2007 Framework Agreement, clause 6. See also Appendix B, JCT 2005 communications.
[911] JCT 2007 Framework Agreement, clause 10.2.

## Information

Another aspiration is the requirement to share 'knowledge' where it would be of value to the other party in the performance of the Underlying Contracts or would assist in the performance of the projects to which those Underlying Contracts relate[912]. While the parties are expected to 'promptly volunteer and share such knowledge or information', there is no provision for any reward or for the creation of a preconstruction phase arrangement during which such sharing would be of most value[913]. Nor is there any provision describing the intellectual property rights to which such information and know-how would need to be subject.

## Conclusion

In conclusion, the JCT 2007 Framework Agreement represents recognition by the JCT that collaborative working processes can have a place in the contractual structure. However, by retaining the twin-track approach of putting the collaborative processes in one document (the JCT 2007 Framework Agreement) and a series of overriding operational provisions in other documents (the Underlying Contracts), the JCT potentially create more problems than they solve.

# *NEC3 framework contract*

As is typical of the NEC suite of contracts, the NEC3 Framework Contract is drafted in very straightforward language and comprises very few provisions – one and a half pages of 'Core Clauses' and a further one and a half pages of 'Contract Data' for completion with details of the 'Employer', the 'Supplier', the 'Framework Information' and procedures governing the framework relationship[914].

## Procedures

The NEC3 Framework Contract envisages a system for the award of successive project contracts. Its Contract Data includes space for insertion of 'the scope', 'the selection procedure' and 'the quotation

---

[912] JCT 2007 Framework Agreement, clause 11.
[913] JCT 2007 Framework Agreement, clause 11.1.
[914] NEC3 Framework Contract.

*Appendix C*

procedure'[915]. This allows the parties to insert a description of a particular type of project or programme of works or services, and the procedures for award of successive projects and the creation of successive project-specific contracts.

However, that is as far as the NEC3 Framework Contract goes. The parties are left to devise for themselves the provisions whereby the appointed contractor is selected for a particular project and whereby the contractor quotes for that project. The NEC3 Framework Contract does not offer any models or options to populate the required selection and quotation procedures, so that there is no equivalent to, for example, the pricing options that appear in NEC3.

## Work packages

One anomaly in the NEC3 Framework Contract is that it provides for submission of a quotation in respect of the proposed 'Work Package' only after the client has selected the contractor[916]. This appears to exclude the possibility of a 'mini-competition' between alternative contractors under equivalent framework agreements, an approach specifically contemplated by the Public Contracts Regulations and one that many clients will rely on as a means to establish best value in respect of each project.

An interesting feature of the NEC3 Framework Contract is that it contemplates the issue to the contractor of a 'Time Charge Order', which comprises 'an instruction to provide advice on a proposed Work Package on a time charge basis'[917]. This suggests that there should be a contractual basis upon which the contractor could undertake a range of preconstruction phase activities in order to assist in finalising details of the proposed project. However, the NEC3 Framework Agreement does not provide guidance detailing any preconstruction processes that might be covered by such a Time Charge Order.

## Other NEC contracts

The NEC3 Framework Contract and its Time Charge Orders do not function as stand-alone documents. The NEC3 Framework Contract Guide makes it clear that the client should at the same time enter into an NEC3 Professional Services Contract to govern work under a Time Charge Order, namely a second parallel contractual arrangement[918].

---

[915] NEC3 Framework Contract, Contract Data Part One – Data provided by the Employer.
[916] NEC3 Framework Contract, clause 22.1.
[917] NEC3 Framework Contract, clause 11.2(5).
[918] NEC3 Framework Guide, 2.

This is a cumbersome and potentially confusing basis upon which to commission preconstruction phase activities.

## Option X12

The NEC could have further developed the NEC3 Framework Contract by incorporating collaborative provisions equivalent to those set out in NEC Option X12. For example, it could have included provision for:

(1) A framework core group identifying representatives of the client and contractor responsible for operating the framework relationship within specific terms of reference;
(2) Early warning at framework level in respect of problems encountered by either party;
(3) Measures by way of key performance indicators and agreed targets, to determine whether the framework relationship is working and to allow the framework core group to assess progress and implement action plans to overcome obstacles or difficulties.

The NEC3 Framework Contract contains none of these. Works or services are to be called off under whatever selection and quotation procedures may be agreed, and any of the above provisions would need to be incorporated in these procedures or in separate bespoke drafting.

## Duration

In order to commit to a framework arrangement, the client and the contractor should be able to rely on each other operating a framework contract for a minimum period of time, subject to the parties meeting their agreed performance targets. However, the NEC3 Framework Contract allows either party to 'terminate their obligations under this Contract at any time'[919]. This does not offer a stable commercial basis for the increased investment and commitment that framework arrangements are intended to attract.

## Problem solving

Finally, the NEC3 Framework Contract is silent on the subject of problem solving and dispute resolution. This is hard to explain. The

---

[919] NEC3 Framework Contract, clause 90.1.

*Appendix C*

NEC3 Framework Contract includes obligations on both the client and the contractor to operate agreed selection and quotation procedures and, in the event of a breach, there is the possibility of dispute. A framework core group could have been a valuable forum to try to head off such a dispute before it undermines working relationships. Also, provision for structured negotiation or mediation or conciliation would have been worthwhile, so that the long-term nature of the framework commitment can be sustained without the rupture that is likely to occur when either party resorts to legal proceedings.

## APPENDIX D
# FORM OF RISK REGISTER
## See Chapter 4, Section 4.4.4 (Risk management and contracts)

(taken from PPC2000[920])

| Risk | Likelihood of Risk | Impact of Risk on Project | Partnering Team member(s) responsible for Risk Management | Risk Management Action | Action Period/Deadline |
|---|---|---|---|---|---|
|  |  |  |  |  |  |
|  |  |  |  |  |  |
|  |  |  |  |  |  |
|  |  |  |  |  |  |

*Guidance Note: The Risk Register should state clearly the nature of each risk, its likelihood and impact on the Project (including any anticipated financial impact and proposed risk contingency), the Partnering Team member(s) responsible for Risk Management actions, the agreed Risk Management actions (including actions to reduce the likelihood of each risk and to reduce its financial and other impact) and the agreed periods/deadlines for such actions.*

*Risk Management actions and periods/deadlines should be integrated with the Partnering Timetable and, to the extent that further Risk Management actions are agreed to be undertaken after signature of the Commencement Agreement, should be integrated with the Project Timetable.*

*Risk Management actions should meet the requirements of clause 12.9 of the Partnering Terms.*

[920] This Form of Risk Register appears in PPC2000 Appendix 7. © Association of Consultant Architects Limited and Trowers & Hamlins LLP 2008, see www.ppc2000.co.uk for further information.

## APPENDIX E
# FORM OF PARTNERING TIMETABLE
## See Chapter 5, Section 5.4.2 (Preconstruction phase programmes)

(taken from PPC2000[921])

| Description of Activity/Requirement | Clause of Partnering Terms (if relevant) | Partnering Team Member(s) Responsible for Activity | Period/Deadline for Activity | Additional Comments |
|---|---|---|---|---|
|  |  |  |  |  |
|  |  |  |  |  |
|  |  |  |  |  |

*Guidance Note: The Partnering Timetable should state clearly the nature, sequence and duration of the agreed activities of each Partnering Team member and should identify any requirements (whether from Partnering Team members or third parties) that are preconditions to any subsequent activities, in each case throughout the period until the proposed signature of the Commencement Agreement.*

*Activities identified in the Partnering Timetable should include without limitation design development submissions (clause 8.3), surveys and investigations (clause 8.4), updated cost estimates (clause 8.7), Value Engineering (clause 8.8), Business Case submissions (clause 10.3), Specialist tenders (clause 10.6), Risk Management actions (clauses 12.9 and 18.1 and any Risk Register), Client approvals/comments in response to each of the foregoing, and other activities required for satisfaction of preconditions to implementation of the Project on Site (clause 14.1) plus as scheduled Core Group meetings, Partnering Team meetings and Design Team meetings (causes 3.5, 3.8 and 8.13), workshops and other activities to be organised by the Client Representative (clause 5.1).*

[921] This Form of Partnering Timetable appears in PPC2000 Appendix 6. © Association of Consultant Architects Limited and Trowers & Hamlins LLP 2008, see www.ppc2000.co.uk for further information.

# APPENDIX F
# ASSOCIATION OF PARTNERING ADVISERS CODE OF CONDUCT[922]

See Chapter 9, Section 9.6 (The role of the partnering adviser).

A member of the Association shall at all times exercise reasonable skill, care and diligence in the practice of a Partnering Adviser's duties.

A member of the Association shall faithfully and diligently carry out any duties undertaken as a Partnering Adviser, having proper regard to the interests of (1) the client, (2) those persons who will be engaged in the design, construction and maintenance of the project and (3) other consultants.

A member of the Association may not accept an appointment as a Partnering Adviser if there is reason to believe that adequate resources and experienced personnel will not be available to execute the appointment.

A member of the Association shall, before or on entering into formal agreement on an appointment, set out in writing the terms of the appointment including:

- The scope of the services to be provided by the member;
- The responsibilities of the member and the limit of the member's liability;
- The method of calculation of, and timing of payment of, the member's remuneration pursuant to the appointment; and
- Provisions for the termination of the appointment.

A member of the Association may not assign any part or all of an appointment as a Partnering Adviser without the written approval of the client.

A member of the Association shall at all times act impartially as a Partnering Adviser and shall fully cooperate with such other parties to the project and pay due regard to the statutory obligations and qualifications of all other parties associated with the project.

A member of the Association shall fully disclose to any prospective client any existing or potential conflict of interest which might give rise to doubts as to the member's ability and integrity to make independent judgements as a Partnering Adviser during the course of the project.

*Appendix F*

A member of the Association shall not take discounts, commissions or gifts from, or show favour to, any person or bodies associated with any appointment as a Partnering Adviser.

A member of the Association shall not, following a competition or tender for or in respect of an appointment as a Partnering Adviser, either (1) attempt to supplant any other member of the Association in the project to which the competition or tender relates or (2) offer an amended fee known to be lower than any fee submitted by another member of the Association in the competition or tender.

If a member of the Association is required to submit a tender to a client or to a prospective client for or in respect of an appointment as a Partnering Adviser, that member shall not, in the preparation, amendment or finalisation of the tender, make use of any information obtained by that member in an official capacity regarding any other tender.

A member of the Association shall not employ or engage any person as a Partnering Adviser who has been suspended or expelled from membership of the Association (unless that person has been re-admitted as a member of the Association).

A member of the Association shall ensure, in making the member's availability and expertise as a Partnering Adviser known by whatever means, that any information given is factual and relevant and is in no way unfair to competitors and that no information or statement is given or made by the member which could bring the Association or its members into disrepute.

---

[922] See www.partneringadvisers.co.uk for further information.

# APPENDIX G
# BIBLIOGRAPHY

## Part 1   Contract forms

ACE (2002): Association of Consulting Engineers Agreement A(2) 2002. Mechanical & Electrical Engineering Services Lead Consultant, Association of Consulting Engineers (revised 2004), London.

APM (1998): Association for Project Management Standard Terms for the Appointment of a Project Manager. The APM Group Ltd, Norwich.

CIC (2007): CIC Consultants Contract Conditions: Conditions of Contract for the Appointment of Consultants on Major Building Projects: CIC/Conditions (first edition), RIBA, London.

GC/Works/1 (1999): GC/Works/1 Contract for Building Civil Engineering Major Works. The Stationery Office, London.

GC/Works (2000): GC/Works/1, Amendment 1: applicable to GC/Works/1 Single Stage Design and Build (1998) and GC/Works/1 Two Stage Design and Build (1999).The Stationery Office, London.

GC/Works/1 Two Stage Design and Build (1999): GC/Works/1 Two Stage Design and Build. General Conditions. The Stationery Office, London.

JCT 2005 Design and Build (2005): JCT 2005 Design and Build Contract (DB). Sweet & Maxwell, London.

JCT 2005 Framework Agreement (2005): JCT 2005 Framework Agreement (FA). Sweet & Maxwell, London.

JCT 2007 Framework Agreement (2007): JCT 2007 Framework Agreement (FA). Sweet & Maxwell, London.

JCT 2005 MP (2005): JCT 2005 Major Project Construction Contract (MP). Sweet & Maxwell, London.

JCT 2005 Minor Works (2005): JCT 2005 Minor Works Building Contract (MW). Sweet & Maxwell, London.

JCT 2005 SBC/AQ (2005): JCT 2005 Standard Building Contract With Approximate Quantities (SBC/AQ). Sweet & Maxwell, London.

JCT 2005 SBC/Q (2005): JCT 2005 Standard Building Contract With Quantities (SBC/Q). Sweet & Maxwell, London.

JCT 2005 SBC/XQ (2005): JCT 2005 Standard Building Contract Without Quantities (SBC/XQ). Sweet & Maxwell, London.

JCT 1998 (1998): JCT 1998 Private With Quantities. RIBA Publications, London.

JCT CA (2008): JCT 2008 Consultancy Agreement (Public Sector) (CA). Sweet & Maxwell, London.

JCT CE (2006): JCT 2006 Constructing Excellence Contract (CE). Sweet & Maxwell, London.
JCT CE Project Team Agreement (2008): JCT 2008 Constructing Excellence Contract, Project Team Agreement (CW/P). Sweet & Maxwell, London.
JCT CM (2008): JCT 2008 Construction Management Agreement (CM/A) and JCT 2008 Construction Management: Trade Contract (CM/TC). Sweet & Maxwell, London.
JCT Management Contract (1986): JCT 1986 Management Contract. RIBA Publications, London.
JCT PCSA (2008): JCT 2008 Pre-Construction Services Agreement (General Contractor) (PCSA). Sweet & Maxwell, London.
JCT PCSA(SP) (2008): JCT 2008 Pre-Construction Services Agreement (Specialist) (PCSA/SP). Sweet & Maxwell, London.
JCT WCD (1981): JCT 1981 Standard Form of Building With Contractor's Design edition. RIBA Publications, London.
LIFT (2006): LIFT Lease Plus Agreement. Community Health Partnerships, London.
NEC2 (1995): Engineering and Construction Contract; an NEC document, second edition. Thomas Telford, London.
NEC3 (2005): Engineering and Construction Contract, third edition. Thomas Telford, London.
NEC3 Framework Contract (2005): NEC3 Framework Contract, first edition. Thomas Telford, London.
NEC3 Professional Services Contract (2005): NEC3 Professional Services Contract, third edition. Thomas Telford, London.
NEC3 Short Contract (2005): NEC3 Engineering and Construction Short Contract, first edition. Thomas Telford, London.
Perform 21 (2004): Perform 21 Federation of Property Societies, Public Sector Partnering Contract. BLISS, Warrington.
Perform 21 PSPCP (2004): Perform 21 Public Sector Partnering Contract Partnering Agreement. BLISS, Warrington.
Perform 21 PSPC 3 (2004): Perform 21 Public Sector Partnering Contract Authority Design Lump Sum Contract. BLISS, Warrington.
Perform 21 PSPC 9 (2004): Perform 21 Public Sector Partnering Contract Professional Services Contract. BLISS, Warrington.
Perform 21 PSPC 10 (2004): Perform 21 Prestart Agreement. BLISS, Warrington.
PfS (2008): Partnerships for Schools: Building Schools for the Future. Standard Form Strategic Partnering Agreement. Report prepared for PfS by Bevan Brittan and Addleshaw Goddard. PfS, London.
PPC2000 (2008): PPC2000 ACA Form of Contract for Project Partnering 2000 (amended 2008). The Association of Consultant Architects, Kent.
RIBA (2004): Royal Institute of British Architects, Standard Form of Agreement for the Appointment of an Architect (SFA/99) (updated April 2004), RIBA Publications, London.
RIBA (2007): Royal Institute of British Architects Outline Plan of Works 2007. RIBA, London.
RIBA (2008): Royal Institute of British Architects Plan of Work: Multi-Disciplinary Services. RIBA, London.

RIBA PM (2004): Royal Institute of British Architects, Project Manager for a Construction Project Form of Appointment (PM/99) (updated April 2004). RIBA Publications, London.

SoPC 4 (2007): SoPC4 Standardisation of PFI Contracts Version 4. HM Treasury, The Stationery Office, London.

SPC2000 (2008): SPC2000 ACA Standard Form of Specialist Contract for Project Partnering, 2002 (amended 2008). The Association of Consultant Architects, Kent.

TPC2005 (2008): TPC2005 ACA Standard Form of Contract for Term Partnering, 2005 (amended 2008). The Association of Consultant Architects, Kent.

## Part 2  Articles and books

Arrighetti, A., Bachmann, R. & Deakin, S. (1997) Contract law: social norms and inter-firm cooperation. *Cambridge Journal of Economy*, **21**, 171–195.

Barlow, J., Cohen, M., Jashapara, A. & Simpson, Y. (1997) *Towards Positive Partnering: Revealing the Realities for the Construction Industry.* The Policy Press, Bristol.

Bennett, J. (2000) *Construction – The Third Way.* Butterworths-Heinemann, Oxford.

Bennett, J. & Jayes, S. (1998) *Seven Pillars of Partnering.* Thomas Telford, London.

Bennett, J. & Pearce, S. (2006) *Partnering in the Construction Industry: a Code of Practice for Strategic Collaborative Working.* Butterworth-Heinemann, Oxford.

Berring, G. (1999) *And it Reduces Finance Risks.* Partnering a Primer, The Builder Group Plc (on behalf of the Community Housing Forum), London.

Bresnen, M. & Marshall, N. (2002) Partnering in construction: a critical review of Issues, problems and dilemmas. *Construction Management and Economics*, **18** (2, 1 March), 229–237(9).

British Standard 6079 (2002) *BS 6079-1 Project Management. Guide to Project Management.*

Burke, R. (2002) *Project Management Planning and Control Techniques*, third edition 1999 (reprinted 2002). Wiley, Chichester.

Campbell, D. & Harris, D. (2005) Flexibility in long-term contractual relationships: the role of cooperation. *Lean Construction Journal* (www.leanconstructionjournal.org), **2** (April), 5–29.

Chitty, J. (2008) *Chitty on Contracts*, thirtieth edition. Thomson Reuters (Legal) Limited, London.

Colledge, B. (2005) Relational contracting – creating value beyond the project. *Lean Construction Journal* (www.leanconstructionjournal.org), **2** (1, April), 30–45.

Commons, J. (1934) *Institutional Economics: its Place in Political Economy.* MacMillan Company, New York.

Constructing Excellence website: http://www.constructingexcellence.org.uk

Cox, A. & Thompson, I. (1998) *Contracting for Business Success.* Thomas Telford, London.

Cox, A. & Townsend, M. (1998) *Strategic Procurement in Construction: Towards Better Practice in the Management of Construction Supply Chains.* Thomas Telford, London.

## Appendix G

Duncan-Wallace, I.N. (1996) *Construction Contracts: Principles and Policies in Tort and Contract.* Vol.1 (1986) and Vol. 2 (1996). Sweet & Maxwell, London.

Duncan-Wallace, I.N. (2000) Errors in bull prices or what price partnering? *Construction Law Journal*, **16** (1), 40–43.

Eggleston, B. (2006) *The NEC3 Engineering and Construction Contract: a Commentary*, second edition. Blackwell Publishing, Oxford.

Forward, F. (2002) *The NEC Compared and Contrasted.* Thomas Telford, London.

Greenwood, D. (2001) Sub-contract procurement: are relationships changing? *Construction Management and Economics*, **19**, 5–7.

Hudson, A. (1995) *Hudson's Building and Engineering Contracts*, eleventh edition. Sweet & Maxwell, London.

JCT CE Guide (2006): *JCT – Constructing Excellence Contract Guide.* Sweet & Maxwell, London.

JCT 2005 Framework Agreement Guide (2005): *JCT 2005 Framework Agreement Guide.* Sweet & Maxwell, London.

JCT 2007 Framework Agreement Guide (2007): *JCT 2007 Framework Agreement Guide.* Sweet & Maxwell, London.

JCT Practice Notice 4 (2001): *JCT Standard Forms of Building Contract 1998 editions: Series 2, Practice Note 4, Partnering.* RIBA Publications. London.

Jones, D., Savage, D. & Westgate, R. (2003) *Partnering and Collaborative Working: Law and Industry Practice.* Hammonds LLP, London.

Kumaraswamy, M. (1997) Common categories and causes of construction claims. *Construction Law Journal*, **13** (1), 21–34.

Lane, N. (2005) How to run a seven-year marathon. *Building*, **35**, 54–55.

Latham, M. (2002) Just my opinion. *Building*, **18**, 33.

Latham, M. (2004) The cynic's bestiary. *Building*, **4**, 33.

Latham, M. (2006) Key Note Address to PPC2000 User Group National Conference, 5 October 2006.

Lock, D. (2000) *Project Management*, seventh edition. Gower, Aldershot.

MacNeil, I.R. (1974) The many futures of contracts. *Southern California Law Review*, **47**, 691–816.

MacNeil, I.R. (1978) Contracts: adjustment of long-term economic relations under classical, neoclassical and relational contract law. *Northwestern University Law Review*, **72** (6), 854–905.

MacNeil, I.R. (1981) Economic analysis of contractual relations – its shortfalls and the need for a 'rich classificatory apparatus'. *North Western University Law Review*, **75** (6), 1018–1063.

Milgrom, P. & Roberts, J. (1992) *Economics Organisation and Management.* Prentice-Hall Inc., Englewood Cliffs, NJ.

Mosey, D. (2003) *A Guide to the ACA Project Partnering Contracts PPC2000 and SPC2000.* The Association of Consultant Architects, Kent.

NCA website: http://www.ncahousing.org.uk

NEC Framework Guide (2005): *NEC3 Framework Contract, June 2005 Guidance Notes and Flow Charts.* Thomas Telford, London.

NEC3 Procurement and Contract Strategies (2005): *NEC3 Procurement and Contract Strategies.* Thomas Telford, London.

Nisbet, J. (1993) *Fair and Reasonable – Building Contracts from 1550: a Synopsis.* Stoke Publications, London.

O'Reilly, M. (1995): Risk, construction contracts and construction disputes. *Construction Law Journal*, **11** (5), 343–354.

PPC Pricing Guide (2008): *Introduction to Pricing under PPC2000 for use with ACA Project Partnering Contracts PPC2000 and PPC International*. The Association of Consultant Architects Limited, Kent

PPC2000 website: http://www.ppc2000.co.uk

Rawlinson, S. (2008) Procurement: single stage tendering. *Building*, **46**, 68–69.

Rhys-Jones, S. (1994) How constructive is construction law? *Construction Law Journal*, **10** (1), 28–38.

RIBA (1999): *Royal Institute of British Architects –Architects' Appointments, Notes for Architects on the Use and Completion of SFA/99*. RIBA Publications, London.

RIBA Guide (2004): *Architect's Contract Guide to RIBA Forms of Appointment, Revisions*, second edition. RIBA Publications, London.

Selznick, P. (1969) *Law, Society and Industrial Justice*. Russell Sage Foundation, New York.

Skeggs, C. (2003) Project partnering in the international construction industry. *International Construction Law Review*, **20** (4), 456–482.

Smith, N.J. (2002) *Engineering Project Management*, second edition, Blackwell, Oxford.

Smith, N.J. (2003) *Appraisal, Risk and Uncertainty*. Thomas Telford, London.

Smith, N.J., Merna, T. & Jobling, P. (2006) *Managing Risk in Construction Projects*, second edition, Blackwell, Oxford.

Smith, R.J. (1995) Risk identification and allocation: saving money by improving contracts and contracting practice. *International Construction Law Review*, **12** (1), 40–71.

Thornton, Judge Anthony QC (2004) *Ethics and Construction Law: Where to Start?* TECBAR Review, Society of Construction Law, London, April 2004, 23 pp.

Wates website: http://www.wates.co.uk

Williamson, O.E. (1979) Transaction – cost economics: the governance of contractual relations. *Journal of Law & Economics*, **22** (2), 233.

Williamson, O.E. (1985) *The Economic Institutions of Capitalism*. Collier Macmillan, New York.

Williamson, O.E. (1993) The evolving science of organisations. *Journal of Institutional and Theoretical Economics (JITE)*, **149** (1), 36–63.

# Part 3    Government/industry reports

AE (1999): *Constructing the Best Government Client*. Achieving Excellence, HM Treasury London.

Arup (2008): *Partnering Contract Review* for Office of Government Commerce. Arup, London, 25 September 2008.

Banwell Report (1964): *The Placing and Management of Contracts for Building and Civil Engineering Work*. Banwell Report, HMSO, London.

Be PFI (2003): *Improving PFI Through Collaborative Working*. Be-Collaborating for the Built Environment, Reading.

CABE (2006): *HM Government Better Public Building*. Commission for Architecture and the Built Environment, London.

## Appendix G

CE (2004): *Demonstrating Excellence – an Evolution of the Programme of Demonstrations*. Constructing Excellence, London.

CE (2006): *Procurement Costs in the Construction Industry – a Guide for Clients, Consultants and Contractors*. School of Construction Management and Engineering, University of Reading, Constructing Excellence in the Built Environment, London.

CIC (2002): *Guide to Project Team Partnering*. Construction Industry Council, London.

CIRIA (1998): *Selecting Contractors by Value*. Construction Industry Research and Information Association, London.

Efficiency Unit (1995): *Construction Procurement by Government – an Efficiency Unit Scrutiny*. Efficiency Unit, Cabinet Office, HMSO, London.

Egan, J. (1998) *Rethinking Construction*. Report of the Construction Task Force, published by Department of the Environment, Transport and Regions, London.

Egan, J. (2008) Transcript of Sir John Egan's speech at the House of Commons, marking the tenth anniversary of *Rethinking Construction* (21 May).

Emmerson, H. (1962) *Survey of Problems Before the Construction Industries*. Ministry of Works, HMSO, London.

Highways Agency (2005): *Highways Agency Procurement Strategy Review 2005*: http://www.highways.gov.uk

Housing Forum (2000): *How to Survive Partnering – It Won't Bite*. Report of the Housing Forum, Partnering working group, London.

ICE/DETR (2001): *Managing Geotechnical Risk: Improving Productivity in UK Building and Construction*. Prepared under the DETR Partners in Technology Programme for the Institution of Civil Engineers, Thomas Telford, London.

Kier (2005): A PPP in Housing/Building Repairs and Maintenance, 4ps. Kier Group and Sheffield City Council, project information briefing.

Latham, M. (1994) *Constructing the Team*. Final Report of the Government/Industry Review of Procurement and Contractual Arrangements in the UK Construction Industry, HMSO, London.

MoD (2007): *A Partnering Handbook for Acquisition Teams*. Ministry of Defence, Defence Commercial Directorate.

NAO (2001): *Modernising Construction*. Report by Comptroller and Auditor General, The Efficiency Office, The Stationery Office, London.

NAO (2005): *Improving Public Services through Better Construction*. National Audit Office, The Stationery Office, London.

NAO (2008): House of Commons Committee of Public Accounts (2008) *The Roll-out of the Jobcentre Plus Office Network*. The Stationery Office, London.

NEDC (1991): *Partnering: Contracting Without Conflict*. National Economic Development Council Construction Industry Sector Group, HMSO, London.

NEDO (1975): *The Public Client and the Construction Industry*. National Economic Development Office, HMSO, London.

Nichols (2007): *Review of Highways Agency's Major Roads Programme*. Nichols Group, London.

ODPM/LGA (2003): *National Procurement Strategy for Local Government ODPM/Local Government Association*. ODPM/LGA, London.

OGC (2007): *Achieving Excellence in Construction*. Construction Projects Pocketbook. Office of Government Commerce, London and *Achieving Excellence in Construction Procurement Guides*, Nos 1–11, Office of Government Commerce, London.

RICS (2001): *Contracts in Use; a Survey of Building Contracts in Use in 2001*. Royal Institution of Chartered Surveyors, London.

RICS (2004): *Contracts in Use: a Survey of Building Contracts in Use in 2004*. Royal Institution of Chartered Surveyors, London.

SCL Protocol (2002): *Society of Construction Law Delay and Disruption Protocol*. Society of Construction Law, Oxford.

Strategic Forum (2001): *Report on Demonstration Projects*. Strategic Forum, London.

Strategic Forum (2002): *Accelerating Change*. Strategic Forum for Construction, London.

Trust & Money (1993): *Trust and Money – Interim Report of the Joint Government/Industry Review of Procurement and Contractual Arrangements in the United Kingdom Construction Industry*. HMSO, London.

# Part 4  Table of cases

Baird Textiles Holdings Limited v. Marks & Spencer Plc [2001] EWCA Civ 274.

Birse Construction Limited v. St David Limited (No.1) [2000] B.L.R.57.

Costain Limited and Others v. Bechtel Limited [2005] EWHC 1018 (TCC).

Courtney & Fairbairn Limited v. Tolaini Bros (Hotels) Limited [1975] 1 W.L.R. 297.

Henry Boot Construction Limited v. Alsthom Combined Cycles Ltd [1999] Build L.R. 123.

Tesco Stores Limited v. The Norman Hitchcox Partnership Limited & Others [1997], 56 Con. L.R. 42.

Yorkshire Water Authority v. Sir Alfred McAlpine & Son (Northern) Limited [1985] 32 B.L.R.114.

# Index

Arup report
    on contracts as procurement systems, 13
    on core group, 108
    on early contractor appointment, 58–9, 206
    on letters of intent, 132
    on managing cost of change, 75
    on partnering adviser, 184
    on subcontractor and supplier selection, 90–1
    on two-stage pricing, 23
Association of Consultant Architects
    and Association of Partnering Advisers, 196
    introduction to pricing under PPC2000, 200
Association of Consultants and Engineers
    and early contractor involvement, 63
    on programmes, 117
Association of Partnering Advisers
    code of conduct, 300–301
Association of Project Managers
    appointment of project manager, 155, 157–8
    on programming, 157–8
    and services of project manager, 153

Banwell
    on changes to letting of contracts, 137
    on client involvement, 96
    on contractor design contributions, 59–61
    on early contractor involvement, 57
    on partnering, 161
    on programmes, 112
    on separation of design and construction, 6

on two-stage pricing, 71–4, 76
behaviour
    and alignment of interests, 37
    commercial influences on, 25
    cooperative, 35, 37, 148 – 9
    and frameworks, 146–9
    and long-term relations, 36
    opportunistic, 33–4, 216
    and partnering, 165
Building Schools for the Future
    and early contractor appointments, 146, 207–8
    and frameworks, 145–6, 207–8

case studies
    Constructing Excellence, 209
    National Audit Office, 12–14, 62, 88, 199, 200
    NEDO, 11
    Strategic Forum, 208
    on use of framework agreements, 251–261
    on use of preconstruction phase agreements, 225–250
claims
    and building contracts, 51–5, 216
    causes of, 46–51
    and late contractor appointments, 7
    and preconstruction phase activities, 48–51
client
    need for involvement of, 95–8
    and preconstruction phase processes, 100–101
    and project manager, 152
    role of, 16, 95–8, 218–9

role under standard form building
contracts, 98–100
slow response by, 48
under multi-party contracts, 100
under two-party contracts, 99, 100
communication
between individuals, 102–3, 218
by core group, 106–8, 218
by early warning, 108–11
by notices and meetings, 103–6
poor, 49
role of, 17, 102–11, 218–9
and teamwork, 109
under standard form building contracts, 103–4
conditional contracts
case studies of, 225–261
and choices, 31
concerns as to, 185–6
and consideration, 34–5
effect of number of parties on, 27
and unknown items, 32
Constructing Excellence
demonstration projects of, 209
industry membership of, 201
Construction Industry Council
consultant appointments, 63
on multi-party contracts, 40
on partnering adviser, 195
on partnering charters, 133, 174
on partnering contracts, 3, 170
Construction Industry Research and Information Association
on contracts governing early contractor appointment, 137–8
on contractor design contributions, 198
on partnering, 161
on payment for early contractor involvement, 12
on selecting contractors by value, 61, 72, 200–201
on two-stage pricing, 72, 75–6
construction industry support
for best practice bodies, 210
for client involvement, 203–4
for contractor design contributions, 198–9
for early contractor appointments, 222
importance of, 21, 197–8

for joint risk management, 202–3
for selecting contractors by value, 200–202
for two-stage pricing, 199–200
construction management
and early contractor involvement, 178
construction phase building contracts
incompleteness of, 42–3, 216
and lack of trust, 43
limitations of, 15, 16, 41–6, 217
standard forms of, 41–6
typical sequence of activities under Project Flowchart 1, 55
contracts
categories of, 25–6
clarity through, 220
conditional, 22
and conditionality, 31
evolution of, 13–14, 39–40, 216–7
and new procurement systems, 39–40
planning functions of, 29
coordination
and alignment of interests 36–7
multi-party contracts and, 29, 184–5
use of contracts for, 27–9
core group
contractual provision for, 106
decisions of, 107
dispute resolution by, 107–8
meetings of, 107
problem-solving by, 107–8
purpose of, 106
terms of reference for, 106–7

design
components of, 7
contractor contributions to, 15–16, 59–67, 217
deadlines, 115–7
inadequate, 47–9, 53
integration with consultants, 62–3
integration with subcontractors and suppliers, 64–6
and lead designer, 59–60
payment for contractor, 66–7
design and build
distinguished from joint design, 60–61
and early contractor involvement, 177–8
and risk premiums, 178

## Index

early contractor appointments
  benefits of, 91–2
  case studies of, 225–261
  cost of, 11–12, 32, 66–7, 86, 183–4
  and design, 59–67, 217
  obstacles to, 176–192
  and programmes, 113–5, 118–121
  and project pricing, 8–9, 68–78, 217–8
  and risk management, 78–86, 218
  and risk transfer, 10–11
  and sustainability, 92–4
early warning
  and core group, 107–8
  distinguished from records, 108
  and duty to warn, 108–11
  and project management, 154
  and risk management, 109–111
economic downturn
  and cost certainty, 210–12
  effect on early contractor involvement of, 209–13
  and origins of Constructing the Team, 214–5
Eden Project
  case study, 257–9
  and frameworks, 143
  and risk sharing, 82
education and training
  benefits of, 193–4
  lack of, 194
  need for, 188
Egan, Sir John
  on client dissatisfaction, 46
  on client leadership, 99
  on cost of contracts, 3
  on early appointment of subcontractors, 87
  on integrated project teams, 223
  on negative effect of building contracts, 135
  on partnering, 204
  on PPC2000, 263
  on success of Rethinking Construction, 222–3

framework agreements
  attraction of, 139–141
  and behaviour, 146–9
  case studies of, 251–261
  forms of, 143–5
  and partnering, 141–2
  preconstruction commitments under, 18–9
  preconstruction phase processes under, 142–3, 220
  and private finance initiative, 145–6

GC/Works/1
  communication under, 275
  design development under, 263–5
  early contractor involvement under, 126
  integration of team under, 288
  oral instructions under, 104
  partnering under, 171
  programmes under, 283
  risk management under, 271
  two-stage pricing under, 269
  use of, 126
good faith
  duty of, 172–3
government support,
  for best practice bodies, 210
  for client involvement, 203–4
  for contractor design contributions, 198–9
  for early contractor appointments, 222
  importance of, 21, 197–8
  for joint risk management, 202–3
  for selecting contractors by value, 200–202
  for two-stage pricing, 199–200

Highways Agency
  case study, 244–6
  and early contractor involvement, 61, 82, 93, 115, 123, 207
  Nichols Report for, 91–2, 194
  procurement strategy, 207
Housing Forum
  on early contractor appointment, 12
  on partnering, 169–70

incentives
  to achieve better value in future prices, 69–70
  to achieve savings, 75
  to adjust behaviour, 25

JCT CE
  communication under, 282
  design development under, 269
  early contractor appointment under, 129
  integration of team under, 290
  overriding principle, 172
  partnering under, 171
  programmes under, 287–8
  project protocol, 173
  risk management under, 275
  two-stage pricing under, 271
JCT Framework Agreement
  binding or non-binding, 293
  call off under, 29, 143–4
  duration of, 292
  information sharing under, 294
  partnering under, 171, 292
  supply chain under, 293
  and underlying contracts, 143–4, 292–3
JCT 2005
  communication under, 280–2
  design development under, 267–8
  early contractor appointment under, 128–9
  integration of team under, 289–90
  programmes under, 286–7
  risk management under, 274–5
  two-stage pricing under, 270–1
Job Centre Plus project
  case study, 254–6
  and frameworks, 141, 207
  and partnering, 205
  and preconstruction phase processes, 207
  savings under, 207
joint ventures
  and building contracts, 130
  and early contractor involvement, 130

Latham, Sir Michael
  on client dissatisfaction, 46
  on client role, 16, 95
  on contracts, 45, 262–3
  on cynics, 187–8
  on early contractor involvement, 214
  on economic downturn, 212
  on NEC, 45, 262
  on partnering, 204
  on PPC2000, 263
  on role of client, 16

  on separation of design and construction, 6, 62
  on standard form building contracts, 45
  on testing contracts in court, 189
letters of intent, 130–2
  contents of, 131–2
  effect of, 130–1
  limits of, 131–2, 219
  purpose of, 130
  risk under, 131

main contractor
  design contribution of, 15–16, 59–67
  pricing contribution of, 15–16, 68–78
  programming contribution of, 113–5, 118–121
  risk management contribution of, 15–16, 78–86
  sustainability contribution of, 92–4
management contracting
  and early contractor involvement, 178
meetings
  of core group, 107
  cost of, 105
  excuses at, 105
  purpose of, 105–8
Ministry of Defence
  on partnering, 163, 165, 205
motivation
  behaviour and, 28
  influence of contracts on, 28
multi-party contracts
  coordination and, 28, 215
  integration and, 184–5
  partnering and, 40
  PPC2000 as, 184

National Audit Office
  case studies, 12, 14, 62, 88, 199, 200
  on contracts as an incentivising force, 206
  on early appointment of integrated project team, 87–88, 199, 202
  on early supplier involvement, 64
  and improved results, 208–9
  on partnering, 163, 204–5
  and payment for early contractor involvement, 11
  on programmes, 112

*Index*

National Economic Development Office
   on client involvement, 96–8, 203
   on contractor design contributions, 198
   on partnering, 163
   on payment for early contractor
      involvement, 11
   on single-stage pricing, 179–80
   on two-stage pricing, 199–200
NEC3,
   communication under, 275–8
   design development under, 265
   early contractor appointment under,
      126–7
   integration of team under, 288
   and partnering, 171
   programmes under, 283–4
   risk management under, 272
   two-stage pricing under, 269
NEC Framework Contract,
   call off under, 145, 294–5
   duration of, 296
   and partnering, 296
   problem solving under, 296–7
   work packages under, 295
negotiation
   use of planning to minimise, 30
   use of presentation phase agreement to
      minimise, 23–4
non-binding protocols, 132–3
   partnering charters as, 133
   risks of, 132–3, 219
   under JCT CE, 173
   under JCT Practice Note 4, 133
notices
   receipt of, 104
   verbal, 104
   written, 103–4

obstacles to early contractor appointments,
      20–21, 176–196, 221–2
   and education and training, 193
   and industry conservatism, 188–90
   and the partnering adviser, 195
   personal, 186–193
   procedural, 181–186
   project-specific, 176–181
Office of Government Commerce
   on client leadership, 203–4
   on communication, 103
   on early contractor involvement, 206
   on integrated teams, 199
   on joint risk management, 203
   on partnering, 204
   procurement award to Job Centre Plus,
      255
   on selection by value, 202
opportunism
   and late contractor appointments, 216
   and negotiation, 34
   and preconstruction phase agreements,
      34
   risk and fear of, 32–4
overheads
   agreement of
   lump sum or percentage, 75
   separate pricing of, 74–5

partnering
   and behaviour, 165
   benefits of, 162–4
   and building contracts, 168–172, 221
   challenges to, 166–8, 221
   and commercial cooperation, 164–6
   concerns as to, 192–3
   construction industry views on,
      204–5
   and confidentiality, 174–5
   definition of, 160–161
   and estoppel, 173–4
   and framework agreements, 141
   and good faith, 173
   government support for, 204–5
   long-term, 162
   and preconstruction phase, 19–20
   and preconstruction phase agreements,
      20, 160–162, 220–221
   project, 162
   and project management, 24
   roadblocks to, 167
partnering adviser
   Arup report on, 184
   and best practice, 195
   code of conduct of, 196, 300–1
   role of, 195–6
   under PPC2000, 195
partnering charter
   limits of, 133
   risk under, 132–3

313

Perform 21,
  communication under, 279–80
  design development under, 266
  early contractor involvement under, 128
  integration of team under, 289
  partnering under, 171
  programmes under, 285–6
  risk management under, 273
  two-stage pricing under, 270
planning
  to avoid negotiation, 30
  to demonstrate a business case, 30
  transactional, 29–30
  and two-stage agreements, 31
  use of contracts for, 29, 215
power
  bilateral, 33
  unilateral, 33
PPC2000
  communication under, 278–9
  design development under, 265–6
  early contract appointment under, 127–8
  form of partnering timetable, 299
  form of risk register, 298
  integration of team under, 288–9
  introduction to pricing under, 200
  partnering under, 171
  programmes under, 284–5
  risk management under, 273
  two-stage pricing under, 270
  use of, 205–6
preconstruction phase agreement
  and alignment of interests, 36–8
  case studies of, 225–50
  and clarity, 137, 219–220
  conditional, 22–5, 31
  construction industry experience of, 208–9
  cost and time to create, 32,183–4
  and economic downturn, 209–13
  forms of, 125–9
  freestanding, 22, 31
  government views on, 205–8
  and incompleteness, 23–4
  machinery of, 24
  and negotiation, 23–4
  and long-term relations, 35–6
  as neoclassical contract, 26
  as project management tool, 38
  as relational contract, 26
  role of, 18, 57–9, 215–6
  typical sequence of activities under Project Flowchart 2, 56
preconstruction phase processes
  and claims, 48
  and design, 59–68
  neglect of, 8
  and pricing, 68–78
  and programming, 113–5
  and risk management, 78–86
pricing
  accuracy of, 69
  and contractor selection, 70–2
  main contractor contributions to, 68–78, 217–8
  and new information, 69–70
  two-stage, 72–8, 199–200
private finance initiative
  and early contractor involvement, 180–1
  and frameworks, 145–6
process contract
  as hybrid of relational and neoclassical contract, 31–2
  and opportunism, 32
  potential of, 215–6
  preconstruction phase agreement as, 31–2
  as procurement system, 13–14
profit
  lump sum or percentage, 75
  separate pricing of, 74–5
programme
  and consultant designs, 115–7
  contents of construction phase, 119–121
  contents of preconstruction phase, 113–5
  as contract document, 122–3
  early agreement of construction phase, 58, 118–9
  Gantt charts and, 157
  model form of preconstruction phase, 299
  as planning tool, 157, 219
  preconstruction phase, 113–5
  remedies for non-compliance with, 123–4
  and risk management, 118
  risk of failure to agree, 112
  SCL protocol and, 121–2
  and unrealistic targets, 50

*Index*

project management
  definition of, 151
  and integration of team, 155–6
  and preconstruction phase agreements, 159
  and procurement strategy, 153
  and programmes, 156
  purpose of, 151–2
  and risk management, 154
  and role of client, 152–3
project manager
  as representative of client, 156
  impartiality of, 156
  influence of, 19, 220
  role of, 151–6
provisional sums
  pricing of, 70
  and standard form contracts, 42–3
public procurement regulations
  and early contractor appointment, 181–3
  and frameworks, 182–3
  risk of challenge under, 22
  and selection criteria, 183

risk
  allocation, 43–4
  assessment, 52, 58, 80–81
  in ground, 82–3
  and opportunism, 33
  premiums, 10, 178
  reduction, 78
  sharing, 82–4
  transfer, cost of, 10–11
risk management
  and contracts, 84–6
  cost of, 86
  early, 78–9
  ground risk and, 82–3
  main contractor contributions to, 78–86, 218
  and pricing, 10–11
  purpose of, 83
  separate or joint, 79–82, 218
risk register
  model form of, 298
  under JCT CE, 275
  under PPC2000, 273, 298

Royal Institute of British Architects
  appointment of project manager, 154–5, 158
  on client involvement, 101
  on early contractor involvement, 2, 62–3
  on partnering, 169, 205
  on programmes, 117
Royal Institution of Chartered Surveyors
  survey of building contracts in use (2001), 126, 205–6
  survey of building contracts in use (2004), 126, 206

selection by value,
  for early contractor appointment, 8, 72
  of main contractors, 72, 200–202
  of subcontractors and suppliers, 73, 87
single-stage pricing
  attraction of, 4, 179–80
  exploitation of, 44, 249–250
  incomplete information in, 68–71
  in economic downturn, 210–12
  and gambling, 54
  government and, 223
  and lack of trust, 43–4, 179
  results under, 179–80
  and risk premiums, 10
  risks of, 68
  weaknesses of, 4, 8–9, 33, 68
site investigation
  inadequate, 49
  and risk management, 83–4
standard form building contracts
  and complete information, 42–3, 216–7
  criticism of, 45
  evolution of, 13
  origins of, 43–45
  role of, 41–2
subcontractors
  barriers to early appointment of, 88
  competitive pricing by, 8–9, 90
  design contribution of, 63–6, 86–8
  early appointment of, 60, 64, 86–8, 218
  joint selection of, 89–91
  nomination of, 89
  sustainability contribution of, 92–4
suppliers
  barriers to early appointment of, 88
  competitive pricing by, 8–9. 90

design contribution of, 63–6, 86–8
early appointment of, 58, 60, 64, 86–8, 218
joint selection of, 89–91
nomination of, 89
sustainability contribution of, 92–4
sustainability
  and early contractor appointments, 92–4
  and single-stage procurement, 94
  and site waste management plan, 93

tenders
  incomplete information in, 50
  gambling and, 54
  main contract, 52, 58
  subcontract, 52, 58
  time constraints under, 9
trust
  and communication, 26
  and exchange of information, 26, 69
  and influence of contracts, 26–7, 135
  lack of, 43–4

two-stage pricing 72–8
  case studies of, 225–261
  concerns as to, 75–8
  and criteria for main contractor selection, 72
  process of, 70, 72–75
  profit and overheads under, 74–5
  under GC/Works/1, 269
  under PPC2000, 270

unknown items
  and limited efficiency, 32, 216
  and preconstruction phase agreements, 27
  and risk management, 78–86
  and uncontrollable events, 50, 215
unwritten understandings, 134–7
  bad faith and, 134
  and partnering, 135–7
  and reliance on personal relationships, 26, 191–2
  risks of, 134